ROUTLEDGE LIBRARY EDITIONS:
ENVIRONMENTAL POLICY

Volume 10

PROBABILITY IS ALL WE HAVE

PROBABILITY IS ALL WE HAVE

Uncertainties, Delays, and Environmental Policy Making

JAMES K. HAMMITT

Routledge
Taylor & Francis Group

LONDON AND NEW YORK

First published in 1990 by Garland Publishing, Inc.

This edition first published in 2019
by Routledge
2 Park Square, Milton Park, Abingdon, Oxon OX14 4RN

and by Routledge
52 Vanderbilt Avenue, New York, NY 10017

Routledge is an imprint of the Taylor & Francis Group, an informa business

British Library Cataloguing in Publication Data
A catalogue record for this book is available from the British Library

ISBN: 978-0-367-18894-8 (Set)
ISBN: 978-0-429-27423-7 (Set) (ebk)
ISBN: 978-0-367-19029-3 (Volume 10) (hbk)
ISBN: 978-0-429-19992-9 (Volume 10) (ebk)

Publisher's Note
The publisher has gone to great lengths to ensure the quality of this reprint but points out that some imperfections in the original copies may be apparent.

Disclaimer
The publisher has made every effort to trace copyright holders and would welcome correspondence from those they have been unable to trace.

PROBABILITY IS
ALL WE HAVE

Uncertainties, Delays, and
Environmental Policy Making

JAMES K. HAMMITT

Garland Publishing, Inc.
NEW YORK & LONDON 1990

Library of Congress Cataloging-in-Publication Data

Hammitt, James K.
Probability is all we have : uncertainties, delays, and
environmental policy making / James K. Hammitt.
p. cm. — (The Environment—problems and solutions)
Includes bibliographical references.
ISBN 0-8240-0406-X (alk. paper)
1. Environmental policy—decision making. 2. Uncertainty.
I. Title. II. Series.
HC79.E5H327 1990
363.7'056—dc20   90-47223

PRINTED IN THE UNITED STATES OF AMERICA

PREFACE

Environmental quality is a topic of increasing concern. Although the topic attracted attention in earlier eras as well, the last 30 years have witnessed a growing concern about environmental degradation and the threat it poses to human welfare and natural ecosystems. Early in the current wave of interest, issues of local or regional air and water pollution were viewed as most pressing. During the last decade or so, regional and global threats—acid rain, stratospheric ozone depletion, and global warming—have become prominent. Each of these issues poses difficult challenges for policy makers.

Local, regional, or global, environmental issues share a number of attributes that make policy choice difficult. Uncertainties abound. I distinguish between uncertainties about *outcomes*, the physical and social consequences of a policy, and uncertainties about *values*, or social preferences among outcomes. Many environmental issues also involve important dynamic elements that create a dilemma of timing. On one hand, we can expect to learn more over time about the physical and economic processes linking policies to outcomes. By delaying, we may reduce our uncertainty and our chance of choosing ineffective or even counter-productive policies. On the other hand, the physical and biological processes often involve long delays, so the environmental consequences of current activities will persist. At a minimum, this persistence implies that delay may increase the cost of environmental protection; at the extreme, it may make it impossible to prevent serious environmental degradation.

In this book I try to suggest ways to think about environmental-policy choice that respond to the importance of uncertainty and delay. I describe several tools for environmental-policy analysis and illustrate their application to important policy issues. In the first part of the book, dealing with stratospheric-ozone depletion, I describe techniques for accommodating outcome uncertainties. In the second part, which considers the health risks associated with pesticide residues on food, I present methods for limiting value uncertainties. I use the final section to address the issue of potential global climate change, and to describe how the tools developed in the earlier sections can be applied to this new challenge.

This book should be of greatest interest to academic, government, and industry analysts and others concerned with improving methods for environmental-policy making. Readers interested in decision-making methodology more generally, or in the particular environmental issues considered, should also find the book of value. Although the methods described are specifically directed towards environmental issues, I believe the general approach can be extended to other areas of public policy.

ACKNOWLEDGMENTS

My thinking on environmental policy making has been stimulated by interactions with a number of colleagues and teachers. Some provided assistance with substantive research questions; others provided useful comments and reactions. All provided inspiration and encouragement. I specifically wish to recognize colleagues at The RAND Corporation: Jan Acton, John Adams, Arthur Alexander, Jerome Aroesty, Sandra Berry, Frank Camm, Jonathan Cave, James Dertouzos, David Draper, John Haaga, James Hodges, James Hosek, James Kahan, David Kanouse, Emmett Keeler, David Lyon, Bridger Mitchell, William Mooz, Daniel Relles, Elizabeth Rolph, John Rolph, Jeannette Roskamp, and Kathleen Wolf; at Harvard University: Shantayanan Deverajan, Joe Kalt, Robert Klitgaard, David Lax, Herman Leonard, Albert Nichols, Howard Raiffa, Thomas Schelling, James Sebenius, and Richard Zeckhauser; at Lawrence Livermore National Laboratory: Peter Connell and Donald Wuebbles; Anil Bamezai at the World Bank; Toshiya Hayashi at the California Assembly Office of Research; Alan Manne at Stanford University; Timothy Quinn at the Metropolitan Water District of Southern California; and Katsuaki Terasawa at the Naval Postgraduate School.

Much of the work presented here was conducted in collaboration with others and has been published in other forms. Part I, Section A incorporates work published in Hammitt et al. (1986, 1987), Hammitt (1990a), and Camm and Hammitt (1986). Much of Part I, Section B was published as Hammitt (1987). Part II is based on Hammitt (1986) and Hammitt (1990b).

For financial support, I thank the U.S. Environmental Protection Agency, the John M. Olin Foundation, and The RAND Corporation. I particularly thank the RAND Economics and Statistics Department and the Computer Information Systems Department for providing special grants.

- vii -

CONTENTS

FIGURES

TABLES

PROBABILITY IS
ALL WE HAVE

Uncertainties, Delays, and
Environmental Policy Making

CHAPTER 1
OUTCOME AND VALUE UNCERTAINTY IN ENVIRONMENTAL POLICY

Environmental-policy makers are challenged by pervasive uncertainties about the consequences of their decisions. Uncertainties surround both the relationships between candidate policies and the resulting physical outcomes, and the proper valuation of differences in outcomes. The first type of uncertainty I denote "outcome uncertainty." It derives from the complex physical, chemical, and biological processes that determine the effect of changes in human activities on changes in environmental conditions that may either affect human welfare directly or produce further consequences that affect welfare. The second type, "value uncertainty," exists because many of these welfare-affecting changes are not directly traded on economic markets, so their economic valuation is not apparent, and the changes in outcomes may involve unfamiliar and unexplored areas of individuals' preference functions.

The relationships between human activities, environmental conditions, and welfare are continuing. Because understanding of these relationships is likely to improve over time, final policy "solutions" are likely to be inappropriate if not impossible. Instead, environmental policies should be flexible and adapted to the best current understanding of these relationships. A dynamic framework that helps policy makers choose the appropriate policy at each instant, and helps them allocate resources to learning about important relationships, is called for.

This book contributes to the methodology for environmental-policy making through studies of two policy areas, addressing issues of outcome and value uncertainty respectively. Tools are developed to help policy makers understand and confront outcome and value uncertainties. These policy-analytic tools are refined and applied in the context of two case studies: stratospheric ozone depletion and pesticide use on human foods. In the concluding section, the application of the tools to possible global climatic changes due to anthropogenic release of carbon dioxide and other gases (the "greenhouse effect") is examined.

Problems of outcome and value uncertainty are not limited to environmental policy. Similar issues arise in other contexts that involve human health and safety, and in other policy areas. As a result, the tools presented here, or at least the method of structuring policy issues, may be applicable to other areas as well. To take only two disparate examples, in both defense and education the relationships between policy choice and outcomes (e.g., war and its consequences, personal cognitive development) are highly uncertain. The values we would assign to particular changes in outcomes are similarly uncertain. For specificity, I consider only environmental policy questions, concentrating on methods that are appropriate for the most important uncertainties in this area. But many of the ideas presented may be adaptable to other contexts as well.

OUTCOME UNCERTAINTY

To illustrate the importance of outcome uncertainty in environmental-policy choice, consider two archetypal problems: human exposure to toxic chemicals and global or regional ecosystem modification.

Toxic Chemicals

Release of toxic chemicals to the environment can lead to adverse effects on human health. The prototypical causal chain extends from chemical emissions through human exposure to health outcomes. Each step is complex, and elaborate computer-based models may be necessary to incorporate as much of the complexity as is required for reliable description of the outcome.

Chemicals may be released to the general environment through air or water emissions from stationary industrial facilities, from mobile automobile and other internal-combustion engines, from consumer appliances such as gas- or wood-burning stoves, from natural sources such as radon seepage from the earth, and from other sources. Once released, they may be transported throughout the environment and chemically or biologically altered. Accurately summarizing the dispersion and transformation of pollutants through air, water, soil, plants, and animals is a complex task. The details of environmental dispersion may vary over time and with unpredictable local conditions such as weather. Through chemical or biological alteration the emitted substances may be degraded into harmless substances or transformed to more harmful compounds. For example, gasoline-engine emissions can be transformed to noxious ozone and other components of smog, and chlorofluorocarbons (CFCs) can break down in the stratosphere where the chlorine they release may deplete the protective ozone layer. Chemical concentrations may be reduced by dilution in the atmosphere, lakes, or oceans, or increased through bioaccumulation.

Human exposure to the pollutants or their products depends on the distribution of these substances through environmental media and on behavioral patterns. Humans may be exposed through the air they breath, through direct contact, or through ingestion of food, water, or soil. If the pollutants are not uniformly distributed through space and time, human exposure can vary widely depending on the duration and time of day individuals spend in specific locations and on their activities. Exposure to air pollutants may depend on residence and workplace locations, on the time of day the individual is at each location, on when and where he engages in physical exercise, and on other factors. Mathematical models can attempt to capture many of these factors through a microenvironment approach (Duan, 1985) or personal exposure monitoring techniques may be applicable (Wallace, 1987). The availability of these monitoring techniques depends on the pollutants and exposure routes of concern. Personal monitoring technologies are routinely used for occupational exposure to radiation and are sometimes used for conventional air pollutants, but have not been widely applied to food or water exposures.

Human exposure to certain chemicals may produce a variety of adverse health effects ranging from acute illness or death to chronic effects including various cancers, neurological damage, reproductive effects, and other adverse outcomes. The relationship between chemical exposure and health outcomes is complex, reflecting the route of exposure, size and timing of dose, pharmacokinetic processes that determine the transport and metabolism of the substance in the body, detoxification and DNA-repair mechanisms, synergistic and antagonistic relations among substances, and so forth. Most attention has focused on carcinogenesis. Current theories suggest it is a multistage process where carcinogenic agents may affect one or more stages and may act as initiators, promoters, or both.

For obvious ethical reasons, only limited human experiments are possible. Most information about the relationship between chemical exposure and health outcomes is obtained through epidemiological study and laboratory testing in small mammals, bacteria, and cultured cells. Epidemiological results are difficult to interpret because, as with any natural experiment, it cannot be determined whether contributing factors other than exposure have been adequately controlled. Especially when considering long-latent diseases like cancer, it may be difficult to reliably determine subjects' past exposure to the chemicals of interest and other conditions.

Extrapolation from animal bioassays to human effects has its own severe problems. Typically, two types of extrapolation are required: from high to low dose and from experimental animal to human. The high to low dose extrapolation is required because experimental animals are subjected to much higher doses than humans in order to increase the probability of any adverse effects that may result to levels that are detectable in a sample of a few hundred animals. Typical experimental protocols cannot distinguish an effect from random variation unless the probability of its occurrence is on the order of one chance in ten, but policy makers are often concerned with human risks that may occur on the order of once in one million human lifetimes. Extrapolation over several orders of magnitude is necessarily highly dependent on the mathematical model used, and neither data in the observable range nor biological theory provide adequate justification for selecting from among several of the available models.

Similarly, the extrapolation from experimental animal to human is a source of great uncertainty. As compared with the inbred strains of small mammals typically used in bioassays, humans are approximately 100 times more massive, live about 30 times longer, have slower metabolic rates, are genetically more diverse, are exposed to a far more diverse set of chemical agents, and differ in other relevant dimensions. Conventional interspecies extrapolations assume the probability of a specified health endpoint is the same for different species exposed to chemical doses that bear the same proportion to body mass or to body surface area. (The ratio of body surface area between two species is not measured but is approximated as the ratio of their masses raised to the 2/3 power.) No strong evidence is available for choosing between these rules; the U.S. Food and Drug Administration typically uses body mass and the U.S. Environmental Protection Agency uses surface area (U.S. General Accounting Office, 1987).

Both interspecies and high to low dose extrapolation procedures may be improved with enhanced understanding and modeling of the pharmacokinetic processes that affect substance distribution and metabolism within relevant species, although limits on human experimentation may constrain development. Some progress has been made for selected chemicals (e.g., National Research Council, 1987; Angelo and Pritchard, 1984; Andersen et al., 1987; U.S. Environmental Protection Agency, June 1987, July 1987).

Information on the relationship between exposure and health effects may also be derived through some of the many short-term *in vitro* mutagenicity and other tests that have been developed. Because chronic animal bioassays are expensive and require several years to conduct, short-term tests on bacteria and cultured cells could be valuable for screening the substances to test in bioassays, developing preliminary risk estimates, and other purposes. Lave and Omenn (1986) propose using batteries of short-term *in vitro* tests to estimate carcinogenicity for regulatory purposes, but recent evidence suggests these tests would be of limited use in this role (Tennant et al., 1987).

Ecosystem Modification

A second archetypal environmental-policy problem concerns human activities that substantially alter global or regional ecosystems. Examples include stratospheric ozone depletion, global warming (the "greenhouse" effect), acid deposition ("acid rain"), deforestation, habitat destruction and species extinction. These effects involve displacement of large complex systems, possibly including the entire geosphere and biosphere, from existing equilibria. The relationship between human activities and ecosystem outcomes is uncertain because of the scale and complexity of these systems, which may incorporate important feedback effects and nonlinearities. Prediction of the effects of policies must include consideration of such large scale phenomena as the global carbon cycle, thermal and chemical equilibration between the atmosphere, hydrosphere, and biosphere, and others.

In the case of toxic chemicals, experimentation on humans is generally precluded but experiments on other mammals can be undertaken (although some animal-rights advocates would argue that even animal experiments should not be conducted). For ecosystems, however, temporally or physically isolated experiments may be impossible. The effects of ecosystem perturbations may be irreversible or virtually so on human time scales, effectively increasing the range of plausible outcomes. In many cases, it may not be possible to perturb the environment, observe the response, and then return to the pre-existing conditions if desired. Species driven to extinction cannot be resurrected.

Geographically limited experimentation may be possible for primarily regional effects such as acid deposition, but where the mechanisms involve global phenomena like global warming or ozone depletion, experimentation with the full system necessarily affects the entire planet. Enhanced understanding of these phenomena must rely on natural experiments, as preserved in the paleohistorical record and conditions on other planets, or experimentation with only part of the relevant system. As a result, predicting the effects of alternate polices depends heavily on mathematical modeling of the physical and biological system.

The long time constants characteristic of many of these processes, due to thermal inertia of the oceans, plant and animal life cycles, and other factors, delay recognition that changes are occurring. The system may thus be committed to a substantial change by activities that occur before the possibility of change is even recognized, much less confirmed. The increased atmospheric carbon dioxide resulting from fossil-fuel combustion and deforestation since the industrial revolution may have already committed the earth to a general warming of several degrees centigrade, and past emissions of CFCs may reduce stratospheric ozone concentrations for the next several decades even if emissions are curtailed immediately.

Environmental-policy making does not generally call for once and final decisions. The relationships between human activities, environmental conditions, and human welfare are continuing and understanding of these relationships is likely to improve over time. This suggests that policies should typically be incremental, flexible, and revised as conditions and new information warrant. Moreover, the rate of learning is endogenous; it too can be influenced by policy choices. At one level, government can influence the allocation of research resources to alternative environmental phenomena; at another, the rate at which the current environment is perturbed will affect the magnitude and timing of effects that can be observed. In the extreme case, if a source of perturbation is eliminated we may never learn what effects it would have produced. For example, if CFC production had been abolished in 1974 when the possibility of stratospheric ozone depletion was first recognized, we might never have learned whether the hypothesized depletion could occur.

VALUE UNCERTAINTY

Efficient resource allocation across activities requires equating the ratios of marginal benefit to marginal cost. Assuming certain reasonable conditions, such as declining marginal benefits of increased resource commitments, a specified quantity of resources yields the greatest benefit if allocated across activities so that the marginal benefit that can be obtained by committing additional resources to each activity is equalized. If the marginal benefit-cost ratios are not equal, total benefits can be increased by reducing the resources allocated to an activity with a relatively low benefit-cost ratio, and adding them to an activity with a relatively high ratio. Resources affected by environmental policies may be allocated more efficiently if methods for valuing the marginal benefits of each policy in a common metric (e.g., dollars) are available.

Economic welfare theory considers only human welfare, which is assumed to depend on satisfaction of individual, fixed and exogenously determined, preferences. These preferences may assign value to the existence of other organisms and of ecosystems in a pristine state, but they need not. Possibly, environmental policy choice should reflect non-human as well as human welfare, but such considerations are not typically incorporated in economic-welfare theory (Stone, 1972; Tribe, 1974; Kelman, 1981).

Environmental policies can affect many kinds of outcomes, including human health risks, resource quality and availability, recreational opportunities, aesthetic qualities, and others that affect human welfare. Determining individuals' preferences for many of these commodities and conditions is often difficult. To the extent these economic goods are traded on economic markets they are inherently tied to other products, so market prices that measure their value directly are not available. Many are public goods, like air quality and other environmental conditions, so other difficulties associated with valuing public goods, such as the commons and free-rider problems, exist. Although the private benefits of many other public goods are conceptually clear and familiar (e.g., the value of the classical common pasture or of irrigation and flood control provided by water projects), many of the benefits associated with environmental policies may involve comparison among regions of individuals' preference functions that are distant from common experience and poorly known, even to the individuals themselves.

Because the effects of environmental policies may extend into the distant future, policy evaluation may require assessment of future preferences. This raises issues of intertemporal and inter-generational preferences and equity. Welfare theory is silent on issues involving changes in population. Current generations may owe some obligation to their descendants, but welfare theory does not incorporate this obligation except to the extent the current generation values bequests it may leave (Broome, 1985; Whittington and MacRae, 1986).

In some sense, a person's identity may change over time; he may be unable to accurately forecast his future preferences and may wish that he had made other decisions in the past. This possibility raises issues of the time consistency of behavior and an individual's ability to constrain his future behavior or to credibly inform others of his future prospects so as to benefit from them in earlier periods. For example, the extent to which an individual values an increase in his life expectancy may depend on whether he can finance current consumption by borrowing against expected future income (Cave, 1988).

An additional factor in valuing future conditions is the possibility of adaptive behavior, technological innovation, and the like. The appropriate comparison is between welfare under each policy after optimizing behavior for the corresponding environmental conditions. Because the nature and extent of innovation and adaptation to unfamiliar conditions is difficult to imagine, the valuation of the difference in conditions is unreliable. Social attitudes and other influences on preferences may also change in ways that are difficult to predict. The valuation of increased skin-cancer risk due to stratospheric ozone depletion might differ significantly between current western society, where a suntan is a status good, and Victorian England, where a creamy complexion was desired.

Because many of the economic goods produced through environmental policy are not traded on markets, people may be reluctant to compare their value to those of marketed goods or to money and may have difficulty in doing so. Even though these goods are not sold directly, people do have experience in choosing actions that affect their holdings of some of them, such as changes in health risks. Through a number of daily decisions, including choice of diet, exercise, using automobile seat-belts, driving when intoxicated, and smoking, people affect the health risks they face. For these goods, assigning a monetary value may be difficult but not impossible. But for environmental conditions that are far removed from common experience, determining one's preferences may be exceedingly difficult. How, for example, should we consider assigning a value to reducing the current rate of species extinction? One approach would consider the instrumental value of existing species to human welfare. This approach might include the value of genetic diversity as a library from which compounds or organisms may be drawn that may have value as medicines or in other direct applications. Or it might consider the role of genetic diversity as a natural laboratory for studying biology, evolution, and other sciences that contribute to understanding of the natural world. Alternatively, many people may feel that humans owe a duty to other life forms to preserve them, although this duty may sometimes be outweighed by other considerations.

Because of the long time periods that may need to be considered, proper discounting of future benefits and costs can be crucial to policy evaluation. Appropriate discounting practice is a confused and contentious area (Lind et al., 1982; Quirk and Terasawa, 1987).

Some commentators argue that values of distinct goods should often be discounted at different rates, and at rates that vary over time (Goodin, 1982). Others argue that any practice except uniform geometric discounting leads to perverse results, such as the conclusion that an action should be taken but delayed indefinitely (Keeler and Cretin, 1983). The most compelling rationale for discounting appears to be the opportunity-cost argument: The value of goods that can be obtained for money should be discounted at the opportunity cost of funds. For goods that cannot be purchased, or when physical irreversibilities are involved, the answer is not as clear and may depend on distributional considerations. Resources that could be used today to cure people afflicted with some disease can be invested and used to cure a larger number of people in the future. Whether this is worthwhile depends in part on how the distributional effect is assessed. But resources that could be used to protect a species from extinction today will not enable some future generation to resurrect it.

Related issues involve the concepts of existence and option values. These involve individuals' valuation of environmental conditions that they may not experience directly, but that they may value either because they may experience them sometime, or because they simply value the knowledge that they exist (Freeman, 1979b, 1984; Desvousges et al., 1987; Smith, 1985).

Valuation Methods

Two classes of methodologies have been developed to measure the benefits of environmental and other benefits, revealed preference and contingent valuation. Revealed-preference methods infer individuals' preferences from their behavior. These methods depend on the existence of either a market good or of some other good with measurable value, for which consumption is related to the value of the environmental good. For example, the value of reduced air pollution has been estimated from differences in housing prices in areas with varying air quality (Harrison and Rubinfeld, 1978), mortality risk has been valued by comparing wages across occupations (Viscusi, 1979, 1983), and sites providing recreational opportunities have been valued by the estimated time and monetary cost of traveling to the site (Clawson and Knetsch, 1966; Feenburg and Mills, 1980).

A necessary condition for developing a benefit estimate using a revealed-preference approach is that the attribute to be valued can be statistically separated from the related commodity. To estimate the value of air quality from differences in housing prices, one must identify regions with differing air quality and statistically control for other differences between houses and neighborhoods. Estimating the value of reducing the health risks of smoking cigarettes might not be possible, if one cannot separate the risks from the benefits of smoking. An intermediate case might be estimating the value of preventing the increased risk of non-melanoma skin cancer that is expected to result from stratospheric ozone depletion. Although exposure to solar radiation, the primary controllable risk factor, is related to several market transactions including occupational choice, sunscreen and protective-clothing purchases, and certain recreational activities, the relationship between solar exposure and skin-cancer risk is confounded with that between exposure and suntans. Possibly, an estimate could be obtained by using tanning-parlor prices to measure the benefits of a suntan, but whether a tan obtained at a tanning parlor and a tan obtained at the beach are close-enough substitutes is not obvious.

In addition, like inferences from any natural experiment, revealed-preference estimates may be subject to selection biases that are not adequately controlled. For example, homeowners in areas with lower air quality are likely to be those who value clean air less, because they are less sensitive to adverse effects or simply because they are less wealthy.

Contingent-valuation estimates are based on people's reported values, solicited by survey. In principle, contingent-valuation estimates of any imaginable change in condition can be obtained, although the reliability may be questioned for conditions far removed from the respondents' experience. Contingent-valuation estimates also suffer from the possibility that respondents may not reply honestly, or may not know their own preferences so actual behavior would differ from the respondents' own expectations. However, contingent-valuation estimates have been derived in a number of environmental-policy contexts and the results have correlated well with estimates from other methods (Cummings et al., 1986a).

OUTLINE

Subsequent chapters are divided among three parts, each concerned with a particular environmental-policy area. Part I develops methods for policy analysis in a context of pervasive outcome uncertainty: potential stratospheric ozone depletion. Within this context, two problems are addressed.

Section A develops a method for assessing and characterizing uncertainty about future emissions of CFCs and other potential ozone depleting substances, using subjective probability distributions. This method could be extended to represent uncertainty about other components of the problem and to integrate across the uncertainties.

Section B develops a decision framework that clarifies the current policy choice, whether to adopt additional emission-limiting regulations now or to delay until we learn more about the relationships between emissions, ozone depletion, and its consequences. This framework allows one to calculate a numerical value that functions like the standard of proof required in judicial proceedings. It clarifies how confident policy makers must be that ozone depletion will occur and present a significant problem in order for them to support imposing additional regulations. The method focuses attention on the most important uncertainties for choosing current policy. It should promote consensus-building by identifying the minimal assumptions on which agreement must be reached in order for policy makers to agree on the appropriate policy choice. Particularly in an area like this, where effective policy may require coordination among many nations, the ability to develop a consensus is vital.[1]

Part II concerns a case with significant value uncertainty: the value of reducing food-borne health risks. It develops and compares estimates using both revealed-preference and contingent-valuation methods for consumer valuation of the risk associated with pesticide residues on food, by comparing conventionally and organically grown produce. Estimates of the incremental value of organically grown produce are consistent across methods but differ significantly between consumers who frequently purchase organic produce and those who purchase only conventional produce. Estimates of the corresponding value of risk reduction are not reliable, however, because of substantial outcome uncertainty about the relative health risks from consuming the two types of produce. Organic-produce consumers believe the risk posed by conventionally grown produce is so much larger than conventional-produce buyers believe that the high incremental value they assign to organically grown produce is not inconsistent with estimates of the value of reducing other health risks reported in the literature.

Parts I and II illustrate some of the difficulties encountered in environmental-policy evaluation because of pervasive outcome and value uncertainty. In addition, techniques for obtaining and incorporating in a policy analysis the best available information about likely outcomes and appropriate values, and the extent of uncertainty about them, are developed and refined. Part III describes how these techniques can be applied to another environmental-policy issue, global warming. The policy-analytic tools developed for the outcome uncertainties in potential ozone depletion appear to be readily transferable. The valuation methods applied to food-borne risks can be used to clarify some of the uncertainties about the valuation of consequences, although the very long lags and pervasive consequences of global warming challenge the welfare-economics paradigm. This discussion of how the tools might be applied to policy analysis for global warming supports the generalizability of these techniques and their applicability to other environmental policy issues.

PART I

OUTCOME UNCERTAINTY:
STRATOSPHERIC OZONE DEPLETION

CHAPTER 2
THE POLICY PROBLEM

Release of chlorofluorocarbons (CFCs) and several related chemicals to the atmosphere may lead to reductions in the concentration of stratospheric ozone and contribute to changes in the earth's climate through the "greenhouse effect." Each effect could threaten existing ecosystems and human welfare: Depletion of stratospheric ozone could increase the quantity of ultraviolet radiation reaching the earth's surface, potentially increasing human skin cancer incidence, promoting cataracts and suppressing immune responses in humans and other animals, and causing other adverse effects to animals, crops and other plants, and valuable materials. The "greenhouse effect" is expected to cause a general global warming that would affect climate and weather patterns, sea levels, agricultural productivity, and the geographic distribution of plant and animal species (see Chapter 18).[1]

Potential ozone depletion is an outstanding example of outcome uncertainty. The central mechanism, the atmospheric transport and chemical processes that may lead to stratospheric ozone depletion, is not well understood. Projections of future depletion are based on complex simulation models that have not been completely reconciled with available measurements of the concentration of ozone and other trace gases. Underscoring this uncertainty, the dramatic spring-time decrease in stratospheric ozone levels over the Antarctic recently discovered (Farman et al., 1985; Stolarski et al., 1986) was not predicted by any of the atmospheric models.

Current and future emissions of CFCs and other potential ozone depleters (PODs) are also uncertain. These gases are emitted as a result of use in a diverse set of consumer products and industrial processes. Current world production and use of these compounds is not routinely measured, but must be estimated from various sources. Emissions must be estimated by modeling the time-pattern of emissions for PODs used in various applications. Similarly, even after conditioning on whatever government policy one wishes to evaluate, future emissions depend on economic and technological factors including demand for POD-using processes and products, and innovation that affects the quantities released to the atmosphere from each application. Emissions occur world-wide, with identical effects on ozone since the gases are well mixed in the atmosphere. Thus, economic and technological factors as well as government policies in all of the major POD-using nations affect future emissions. In addition, the effect of a specified quantity of potential ozone depleters depends on concentrations of other gases, including methane (CH_4), nitrous oxide (N_2O), and carbon dioxide (CO_2), that are emitted through poorly-understood natural as well as human activities.

The effects of stratospheric ozone depletion on the biosphere are perhaps even more uncertain. It is agreed that ozone depletion would allow increased levels of ultraviolet radiation to penetrate to the earth's surface, but the consequences of that increase are highly uncertain. Possible adverse effects include those on human and animal health (skin cancer, cataract formation, immunosuppression), interference with plankton, fish larvae, and other sea-life development, decreased agricultural yields and damage to wild plant life, and damage to plastics, paints and other coatings, and other economically important materials. Except for a few of these possible effects, such as human skin cancer, current understanding is inadequate to do more than speculate on the character of the relevant dose-response functions and likely adaptations to environmental perturbations.

An important feature of the chemicals that may cause ozone depletion is their chemical stability: they do not readily react with others. This stability contributes to their safety and usefulness in a variety of applications but also to their potential threat to stratospheric ozone. Because of this stability most of these chemicals are believed to survive in the atmosphere 50 to 100 years or longer, so their concentrations in the lower atmosphere may remain elevated for many years after release. Only a small fraction of the molecules of these chemicals present in the troposphere (lower atmosphere) is transported to the stratosphere at any time; there the molecules are decomposed by ultraviolet radiation and the freed halogen (chlorine or bromine) atoms may catalytically react with free oxygen and ozone molecules. Consequently, if depletion occurs it may persist for decades, even if emissions are terminated. The radiative effect of these chemicals as greenhouse gases would also be persistent.

Potential ozone depleters differ in the effectiveness with which a unit mass of each destroys ozone because of differences in molecular weight, the number of chlorine or bromine atoms per molecule, and the compounds' atmospheric lifetimes. Table 1 reports estimates of these relative efficiencies and atmospheric lifetimes calculated using the Lawrence Livermore National Laboratory (LLNL) one-dimensional model (Hammitt et al., 1987). Relative depletion efficiencies depend on the quantities of other PODs and trace gases in the atmosphere, but the variation is not large for foreseeable concentrations. The table also illustrates one measure of the PODs' relative potential contributions to ozone depletion, expressed as the estimated share of current emissions weighted by the estimated depletion efficiency per unit mass. Other PODs (including HCFC-22, CFC-114 and 115) are quantitatively unimportant at current and likely future emission rates. Fisher et al. (1990) provide estimates of ozone-depletion efficiencies and atmospheric lifetimes for a number of the compounds currently under investigation as substitutes for the principal PODs. Most of these may also contribute to ozone depletion, although efficiencies are one to two orders of magnitude smaller than those of the principal PODs shown in Table 1.

Whether past and current releases, combined with future releases from already existing products that contain PODs (such as rigid foams, refrigerators, and air conditioning systems), will cause significant depletion and UV increase is not known. Evidence is currently appearing to suggest that past POD emissions have depleted stratospheric ozone, but interpretation of these results is complicated by the relatively limited period for which measurements are available, by changes in UV flux over the solar cycle, by drift in satellite-instrument calibration, and by changes in aerosol content of the atmosphere (which may affect instrument calibration) due in part to the April 1982 eruption of El Chichon (World Meteorological Organization, 1985; Reinsel et al., 1984; 1987; Tiao et al., 1986; Reinsel and Tiao, 1987; Bowman, 1988; Watson and Prather, 1988; Heath, 1988; Watson et al., 1988). A recent consensus report concludes that claims of significant POD-induced depletion (Bowman, 1988; Heath, 1988) cannot be supported because of uncertainty about satellite-instrument calibration, but that reanalysis of existing data and comparison of satellite and ground-based measurements suggests PODs have contributed to the global column ozone loss since 1969. Expected increases in solar flux over the building phase of the solar cycle should offset further POD-induced depletion through 1991 (Watson et al., 1988). No increase in ultraviolet (UV) flux at the surface has been detected (Scotto, et al., 1988), although it has been suggested that this result may be offsetting urban pollution (Grant,

Table 1

ESTIMATED RELATIVE CONTRIBUTIONS TO OZONE DEPLETION

Chemical	Atmospheric Lifetime (Years)	Estimated Global 1985 Emissions (Thousand metric tons)	Relative Depletion Efficiency	Share of Total Contribution (Percent)
CFC-11 (CCl_3F)	76.5	238	1.00	25.8
CFC-12 (CCl_2F_2)	138.8	412	1.00	44.7
Carbon tetrachloride (CCl_4)	67.1	66	1.06	7.6
CFC-113 ($CFCl_2CF_2Cl$)	91.7	138	0.78	11.7
Methyl chloroform (CH_3CCl_3)	8.3	474	0.10	5.1
Halon 1301 (CF_3Br)	100.9	3	11.4	3.7
Halon 1211 (CF_2ClBr)	12.5	3	2.70	0.9
HCFC-22 (CHF_2Cl)	22.0	72	0.05	0.4

Note: Atmospheric lifetime = (steady-state atmospheric burden) / (constant emission rate). Relative depletion efficiency = (O_3 depletion per unit mass of species) / (O_3 depletion per unit mass of CFC-11). Share of total contribution = $100 \ S_i/(\Sigma \ S_i)$ where S_i = 1985 emissions x relative depletion efficiency. Shares do not sum to 100 due to rounding.

Source: Hammitt et al. (1987).

1988). An increase in UV flux has been detected in Switzerland (Blumthaler and Ambach, 1990).

The recently discovered low ozone concentration in the Antarctic springtime (Farman et al., 1985; Stolarski et al., 1986) was not predicted. The explanation for the Antarctic "hole" is thought to include heterogeneous chemical reactions that occur on stratospheric cloud particles and possible transport within the atmosphere. Such clouds are common in the Antarctic, but much less common in the Arctic, springtime, possibly accounting for the absence of a comparable transient depletion in the Arctic springtime (Stolarski et al., 1986; Tung et al., 1986; deZafra et al., 1987; Farmer et al., 1987; Hills et al., 1987; Molina et al., 1987; Solomon et al., 1987; Tolbert et al., 1987; Stolarski, 1988). Similarly, there is some evidence to suggest the anticipated global warming caused by carbon dioxide and PODs may be occurring (see Chapter 18).

Atmospheric models suggest that continued releases at current rates would not cause significant depletion, if current increases in CO_2 and CH_4 concentrations continue (World Meteorological Organization, 1985; Hammitt et al., 1987). (These gases reduce ozone depletion in part by cooling the stratosphere and slowing ozone-destroying reactions there.) However, in the absence of regulation, most observers expect POD emissions to increase (World Meteorological Organization, 1985; Quinn et al., 1986; Hammitt et al., 1987). Under the initial terms of the Montreal Protocol, significant increases are also possible (Office of Technology Assessment, 1988; Wigley, 1988), although current negotiations (summer 1990) may yield more stringent terms. Trends in the atmospheric abundances of other important gases (notably CO_2, CH_4, and N_2O), and even their sources, are not well understood (Chapter 18). Moreover, because current production decisions may affect emissions and ozone concentrations well into the future, it may be wise to limit releases now rather than to await future scientific developments before deciding whether regulations are necessary. Simply stated, by the time our understanding of these phenomena has developed to a state that allows us to confidently assess the relationship between potential-ozone-depleting emissions and environmental consequences, it may be too late to avert significant adverse effects. Whether these adverse consequences would be ultimately reversible is not known; in any event, they are likely to persist for a human generation or more.

CURRENT POD PRODUCTION AND APPLICATIONS

Potential ozone depleters are man-made chemicals. They are released to the atmosphere solely as a consequence of their use in a wide range of industrial processes and consumer products.

Estimates of 1985 use are shown in Table 2. These are reported separately for the United States, the other non-communist countries, and the communist countries. This distinction is relevant because much better information is available for United States than foreign use of most of these chemicals, and for non-communist than communist use. Nearly all of the production of CFC-11 and 12 outside the communist countries is by companies that report their production to the Chemical Manufacturers Association (CMA); production in the communist countries is not reported. Consequently, estimates of world use are based on separate estimates for each of three regions: the United States, the "other reporting countries," and the "communist countries" (including the Soviet Union, Eastern Europe, China, and the other communist Asian nations). The non-communist countries are jointly called the "reporting countries."[2]

Table 2

ESTIMATED 1985 WORLD USE OF
POTENTIAL OZONE DEPLETING SUBSTANCES

(Thousand metric tons)

	World	United States	Other Reporting Countries*	Communist Countries
CFC-11	341.5	75.0	225.0	41.5
CFC-12	443.7	135.0	230.0	78.7
CFC-113	163.2	73.2	85.0	5.0
Methyl chloroform	544.6	270.0	187.6	87.0
Carbon tetrachloride	1,029.0	280.0	590.0	159.0
Halon 1301	10.8	5.4	5.4	0.0
Halon 1211	10.8	2.7	8.1	0.0

Note: * All countries that report production to the Chemical Manufacturers Association excluding the United States.
Source: Hammitt et al. (1986)

As shown by Table 2, the United States currently accounts for an estimated one-quarter to one-half of world use of the seven major PODs. The estimated U.S. share of specific applications of CFC-11 and 12 is generally similar, except for use in aerosols where, because of a late 1970s U.S. ban on CFCs in "nonessential" aerosol applications, the United States accounts for only about 5 percent of world use. Use in the communist countries is estimated to be less than 20 percent of the global totals, although only limited data are available (Hammitt et al., 1986).

The relationship between commercial use and atmospheric release of the potential ozone depleters varies by application. In some uses, such as aerosol propellants, the chemical is inevitably released to the atmosphere as the product is consumed. In others, such as insulating foam and refrigeration equipment, the chemical is contained in a sealed unit and is released only through unintended leakage or after product disposal. The quantities of PODs contained in such products represent a bank that may or may not eventually reach the stratosphere. In any case, the existence of this bank results in a substantial delay between production and release of the chemicals. This delay is on the order of 10 to 20 years for most refrigeration equipment, and may be 50 or 100 years or more for rigid insulating foams.[3]

Primary applications of the seven chemicals include aerosol dispensers, rigid foam insulation and related products, flexible cushioning foams, refrigeration and air conditioning systems, solvents, fire extinguishants, and other products.

CFC-11 and 12 are used in diverse applications. In the CMA reporting countries, CFC-11 is used almost entirely as an aerosol propellant and as a blowing agent in producing rigid and flexible foams. As shown by Fig. 1, these three uses account for an estimated 90 percent of reporting country use. Most of the remainder represents the difference between the sum of estimated CFC-11 use in each product area and total use, based on total production as reported by CMA (1985) for the last several years (Hammitt et al., 1986).

In the United States, CFC use in "nonessential aerosols" was banned by the Environmental Protection Agency (EPA) effective in 1979 (the ban was announced in 1977). As a result, aerosols account for only about 5 percent of U.S. CFC-11 use, whereas rigid foams account for about half. The manufacture of flexible foam accounts for about one-fifth, similar to its share of world use. The unallocated uses, about 18 percent, represent the difference between estimated use in the listed applications and 1985 domestic use estimated from U.S. International Trade Commission production data. Part of the unallocated CFC-11 use is likely to represent refrigeration and air conditioning applications (Hammitt et al., 1986). In addition, it includes storage, packaging, and transport losses that may account for about 2 percent of use (Wolf, 1980).

Estimated reporting country and U.S. use of CFC-12 are shown in Fig. 2. An estimated 32 percent of CFC-12 use in the reporting countries is as an aerosol propellant, whereas 27 percent is used in the refrigeration applications analyzed by Hammitt et al. (1986), including mobile air conditioning, retail food refrigeration, chillers (large commercial and industrial air conditioning systems), and home refrigerators and freezers. The unallocated uses total 22 percent.

In the United States, aerosols account for only about 4 percent of use, about the same share as of CFC-11. The refrigeration applications account for 44 percent, of which mobile air conditioning is by far the largest component. The unallocated uses are about 31 percent of estimated total domestic use. A large part of the unexplained use of CFC-12 may occur in food refrigeration applications, both in the United States and abroad. Again, storage, packaging, and transport losses may account for about 2 percent of total use (Hammitt et al., 1986).

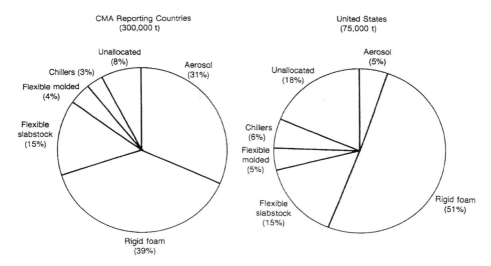

Fig. 1 — Estimated CMA reporting country and United States use of CFC-11, by product.
Source: Hammitt et al. (1986)

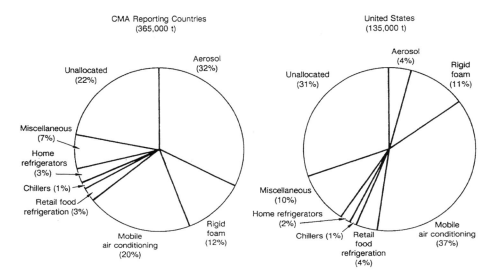

Fig. 2 — Estimated CMA reporting country and United States use of CFC-12, by product.
Source: Hammitt et al. (1986)

The other chemicals are concentrated in fewer applications. Methyl chloroform and CFC-113 are used almost exclusively as solvents. CFC-113 is used largely in the electronics industry to "deflux" printed circuit boards and clean plastic parts including semiconductors; methyl chloroform is a general purpose solvent used in many types of metal and other cleaning applications. Small amounts of CFC-113 are used in dry-cleaning and specialty refrigeration applications.

Carbon tetrachloride is produced in larger quantities than any of the other chemicals but most is transformed into CFC-11 or 12. Additional carbon tetrachloride is used as a solvent, as a grain fumigant, and in the pharmaceutical industry, but chemical producers in the United States agreed to stop using it as an active ingredient in pesticides by 1986.

Halons 1211 and 1301 are relatively new and are produced in small but growing quantities. They are used as fire extinguishants, Halon 1301 primarily in total flooding systems used to protect computers and other valuable equipment in enclosed spaces, Halon 1211 primarily in hand-held extinguishers.

OUTLINE

Uncertainty about the relationship between policies and physical outcomes related to stratospheric ozone depletion has several important implications for policy analysis. In order to evaluate the costs and benefits of alternative policies, it is necessary to account for this uncertainty, and to reflect current understanding of the relationships that are most likely to exist. The uncertainty may itself influence the choice of policy, since a flexible, incremental approach that can be modified as our understanding of these relationships improves may be preferred.

This Part is divided into two sections. Section A develops a convenient, economical method for describing our understanding of the likely range of POD emissions, conditional on a policy of not imposing additional restrictions on POD use.[4] This method produces a set of scenarios for future POD emissions that encapsulate information on the subjective distribution for future ozone depletion. The scenarios bound a subjective probability interval for the joint effect of the seven PODs on model-calculated ozone depletion. In order to fully assess the uncertainties about future ozone depletion and its effects in the biosphere, these scenarios could be combined with similarly developed scenarios describing uncertainty about the modeled atmospheric photochemical and transport processes, and about the relationships between ozone depletion and effects in the biosphere. These scenarios assume no additional regulations are placed on PODs as such, although they do reflect possible regulations on workplace exposures and waste disposal. They provide a baseline against which to compare POD use under alternative policies. To describe the subjective distribution of outcomes corresponding to other policies, the effect of each policy under each of these scenarios could be evaluated.

Section B analyzes the key policy question, recognizing the significant outcome uncertainty and likely improvements in understanding over time. Because of these features, an incremental strategy is likely to be preferred. In this section, the policy dilemma is structured as a stochastic dynamic program with learning. The general formulation is intractable, but a simplified framework is developed that allows calculation of the conditions on policy makers' beliefs about the severity of the problem under which it is better to impose additional restrictions on POD emissions sooner rather than later. These conditions are analogous to the standard of proof required in judicial contexts: They clarify how confident one must be about the relationship between policies and outcomes in order to favor a specific policy.

A. CHARACTERIZING UNCERTAINTY ABOUT OUTCOMES

CHAPTER 3
PROBABILITY-BASED SCENARIOS

Understanding the probability distribution of the outcomes and costs of a policy to restrict POD emissions requires assessment and convolution of uncertainties about a wide range of factors. These include the future emission trajectories of potential ozone depleters and of other gases that influence ozone concentrations, the effect of a specified set of emission paths on actual ozone concentrations, the increase in UV flux to the earth's surface and its effects on human health, crop yields, other aspects of the biosphere, degradation of materials, and other consequences of policy interest. Uncertainties about each of these factors contribute to uncertainty about the magnitude of the potential threat and the effects of alternative global strategies for controlling POD emissions. Moreover, the presence of this pervasive uncertainty is an important factor in choosing an appropriate policy, since a more flexible policy may be preferred to alternatives that would perform better in a more certain environment.

Uncertainty about physical, chemical, biological, and economic effects can be usefully represented using the structure of subjective or Bayesian statistics.[1] Other methods of representing uncertainty have been developed, but because these do not use the probability calculus they can produce inconsistencies (Draper et al., 1987; Lindley, 1987). The most prominent of these non-probability methods are fuzzy set theory (Schmucker, 1984), certainty factors (Barr and Feigenbaum, 1982), and belief functions (Shafer, 1976; 1982).

There are two main alternative interpretations of probability, subjective interpretation and the frequentist interpretation. In the frequentist interpretation, which is the more widely understood of the two, "the probability of event A" is understood to mean the relative frequency of occurrence of event A in some (invariably hypothetical) infinite sequence of repetitions of the mechanism in question. In the subjective interpretation, "the probability of event A" is a representation of one's belief about the likelihood of event A occurring. That is, the first presumes the existence of a stochastic process that can be observed (at least in principle) to gather empirical information about the probability of an event, while the second is a formal way of presenting a subjective judgment. In most if not all public-policy problems, some aspects of the relationship between policies and outcomes are uncertain and not amenable to empirical validation before a policy must be selected. In these and other settings, subjective probability analysis provides a consistent method for incorporating the best available information, including expert judgments, to bear on the problem (Raiffa, 1968; Hodges, 1987).

In order to assess the distribution of possible consequences from a specified policy, it is necessary to understand uncertainty about each of these economic, chemical, physical, and biological factors, and to convolute these uncertainties. To perform this analysis, the problem can usefully be divided into modules and structured as the tree illustrated in Fig. 3. To understand it, start at the top of the figure and work down. Government policies, ranging from bans on CFC use as aerosol propellants to workplace safety standards, currently affect POD emissions. New regulations, such as additional product bans, marketable permits for using PODs, and other policies could also affect emissions. In addition, market and technological factors affect demand for POD-using products and manufacturing processes, and affect the quantities of PODs emitted per unit of product. After release to the atmosphere, PODs and other trace gases diffuse and react in a complex fashion that may be influenced by other factors such as global temperatures and the amount of solar radiation. One of the potential results of these complex interactions is a reduction in the concentration of stratospheric ozone. The diffusion and chemical reactions are simulated by complex computer models that produce time profiles of estimated ozone concentrations at different altitudes and, in some cases, latitudes. These time profiles of ozone concentrations can in turn be transformed into estimated time profiles of increased ultraviolet radiation and the various effects shown at the bottom of the figure, using a variety of models.

Each box has uncertainties associated with it. Moreover, the assessed uncertainty about the outcomes of each box must reflect uncertainties associated with inputs to that box; the uncertainties must be convoluted through the tree. One way to convolute these uncertainties would be to develop models to describe the outputs of each box as stochastic

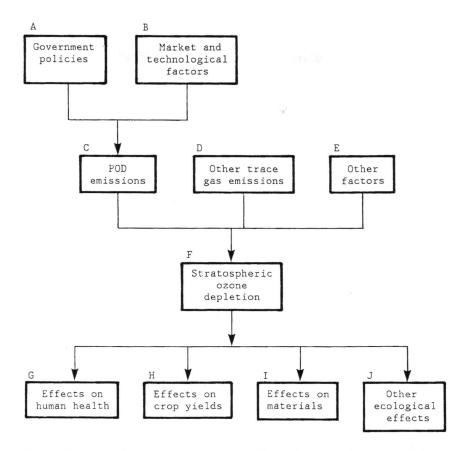

Fig. 3 — Schematic view of potential causes and effects of stratospheric ozone depletion.

functions of the inputs, and then to construct the subjective probability distribution summarizing knowledge and uncertainty about health and ecological effects by Monte Carlo simulation, analytic approximation, or some combination of the two.

Monte Carlo analysis of this problem requires such extensive resources that it does not seem practicable, in large part because of the complexity of the computer-based model used to simulate ozone depletion given specified scenarios for POD and other trace gas emissions. These models typically require several hours to run on modern super-computers. Moreover, so many sources of uncertainty are important that even a simplistic representation, using two or three scenarios to describe the uncertainty about each source, leads to an unwieldy number of cases to use in policy analysis.

However, the structure of the problem lends itself to a simplification that could potentially allow a much less simplistic treatment of uncertainty. Note that all of the information from the top half of the figure funnels through a single box, representing the time profile of stratospheric ozone depletion. This profile is the only input required to study the remaining sources of uncertainty.[2] Thus, in theory, it is possible to develop a subjective probability distribution for the extent of ozone depletion that summarizes all of the uncertainties in the boxes that feed into the potential stratospheric ozone depletion box in Fig. 3 that are relevant to the analysis of the effects in the final row of boxes.

A probability distribution for the extent of ozone depletion could be developed by first developing subjective probability distributions for the uncertain quantities in each of the boxes that feed into the potential stratospheric ozone depletion box and using Monte Carlo analysis to convolute these distributions. Actually, one would want a set of such ozone depletion distributions, each conditional on a specified government policy. These distributions would summarize the extent of understanding of the factors that may influence depletion and would also allow assignment of probabilities to particular ozone depletion scenarios. Thus, one could state that the (subjectively assessed) probability that ozone depletion will fall between two specified levels in a given year is x percent, conditional on the corresponding government policy.

The difficulty with this approach lies in representing the chemical interactions in the atmosphere in a cost-effective way. This difficulty does not eliminate the usefulness of using the concept of a probability distribution for ozone depletion; it can provide a basis for developing emission scenarios used as inputs to the atmospheric models. Ideally, one should be able to interpret cases easily in terms of this subjective probability distribution. For example, a "high" case should be developed from scenarios for market and technological factors, other trace gases, and other factors that together yield a "high" level of ozone depletion, perhaps one consistent with the 75th percentile of the subjective probability distribution for ozone depletion under current government policies. Of course, the subjective probability distribution cannot be accurately estimated without actually running the atmospheric models, which cannot done often enough to generate the distribution. Nonetheless, to the extent that one must choose scenarios for the inputs to the atmospheric models, it makes sense to think about how to choose scenarios in light of what those scenarios are ultimately meant to do: Provide the kind of information that a subjective probability distribution for ozone depletion would provide if it could be developed directly.

SCENARIOS BASED ON A PROXY FOR STRATOSPHERIC OZONE DEPLETION

The scenarios developed here are intended to approximate those that would be derived from a subjective probability distribution for the extent of ozone depletion, if it were to be developed. They describe what I believe to be a reasonable range of POD production and emissions over the period 1985-2040, conditional on the absence of any additional government restrictions on POD use or emissions. Because the factors that are likely to affect POD emissions over the distant future are less clear than those relevant to the near term, the analysis is divided into two periods: 1985-2000 and 2000-2040.

I focus on growth rates for each POD so that a single random variable, the average growth rate over the period, can describe a time profile for production. The main independent sources of uncertainty for the growth rates of of each chemical are identified and designated as "component" random variates, and probability distributions for these variates are parameterized. The available information on economic and technological factors that affect growth in use of these chemicals is embodied in the distributions for these component variates. To the extent that the production growth rates of two chemicals are related, they will both be dependent on at least one common component variate.

The distributions for the individual chemicals are combined using a "score function" that relates production growth rates of the seven PODs to a scalar value that has some policy relevance. The score function is intended to serve as a proxy for some relevant measure of ozone depletion (e.g., the global annual average), so that its subjective distribution can be related to the distribution of ozone depletion. The subjective probability distribution for the score function is derived by convoluting the distributions for each of the component variates underlying the POD production growth rates.

Specific scenarios are selected to represent the range of POD emissions defined in terms of quantiles of the score function. For example, as a a high scenario I use the 95th percentile of the distribution; the subjective probability that the value of the score function will be higher is 5 percent.

Each value of the score function is associated with an infinite number of combinations of component random variates. The production and emissions of each POD associated with these scenarios are the unique set for which all of the components correspond to a common quantile of their respective marginal distributions. This choice distributes responsibility for the joint effect across all seven PODs and their component variates: It associates relatively high- or low-growth scenarios for individual chemicals together. As shown below, the scenario corresponding to the 95th percentile of the distribution of the score function uses the 82nd percentile value for each component variate. Note that the quantiles used for the component variates will differ from the quantile of the score function, and will be closer to their respective medians, unless all of the component variates are perfectly correlated. That is, uncertainty about the calculated joint effect of the seven PODs is proportionally less than uncertainty about the calculated effect of any one of them (conditional on the atmospheric model).

The set of chemical production paths resulting from this process describes a scenario for future production of the seven PODs. In principle, emissions can be generated from it using an extension of this method to reflect uncertainty about the time profile of emissions from each application, and changes in the share of POD production used in each application. At present, however, emissions profiles corresponding to each scenario are generated using a deterministic algorithm. The procedure is explained in more detail below.

Subjective Probability Distributions for Individual Chemicals

The approach begins as a standard Monte Carlo analysis would. A set of independent marginal probability distributions are derived and combined to form a joint probability distribution for the seven PODs. The joint distribution is based on the assumption that all of the correlation among the production of these substances can be represented as uncertainty about general economic growth. Additional production uncertainties are assumed to be independent of general economic growth and of each other. The approach, however, could easily accommodate information on other sources of correlation among the compounds, by adding additional component variables to the analysis.

Score Function

The score function has two roles. First, it provides a scalar value that summarizes information in the subjective joint probability distribution for the individual chemicals. This single value provides a simple way to relate scenarios to one another. Second, the score-function value is designed to have policy significance. Increased production of any of the PODs will increase the value of the score function, and should also be associated with greater ozone depletion. Hence, "high" scenarios should be associated with high ozone depletion and "low" scenarios with low ozone depletion. This relationship between the score function and the extent of ozone depletion is not exact. If it were, the atmospheric models would not be needed. But it is designed to yield scenarios that can be roughly interpreted in terms of the likely corresponding ozone depletion, holding other factors, such as emissions of other trace gases, constant.

One way to think about the score function is as a simple tool that policy makers can use to rank alternative sets of POD production levels. A simple score function would be the sum of POD production in a given year. However, this function would be inadequate because of the widely different effects that a unit mass of each POD may have on stratospheric ozone. The score function proposed here weights the production levels of the chemicals, transforming them to a scale on which a unit of each is believed to have approximately the same effect on the ozone.

The score function chosen is a simple weighted sum of the production growth rates for individual chemicals. The subjectively chosen weights are intended to reflect the chemicals' relative potential to deplete stratospheric ozone. The ith chemical's weight is defined as

$$w_i \equiv (p_i \, f_i \, e_i)/\Sigma_i \, (p_i \, f_i \, e_i)$$

where

p_i = the annual global production of the chemical,

f_i = the fraction of production of the ith chemical that is likely to be released to the atmosphere, and

e_i = the estimated depletion efficiency per unit mass of the chemical relative to CFC-11.

None of these factors can be specified with certainty. Estimates of 1985 production for each chemical are based on the best data available at the time (Hammitt et al., 1986). The fraction of production that will ultimately be released is based on detailed study of the applications of each chemical (Hammitt et al., 1986; Palmer et al., 1980). The estimated relative ozone depletion potencies are based on information from atmospheric models (Hammitt et al., 1987). These relative potencies are sensitive to assumptions in the atmospheric models; they are used only to suggest orders of magnitude for the weights.

Two sets of weights are constructed: one for the period 1985-2000 and another for 2000-2040. The weight for each chemical is proportional to the product of the chemical's annual production at the start of the relevant period and a subjective factor designed to capture the other two terms, f_i and e_i. The factors for CFC-11, 12, and 113 are 1.0 because, despite their diverse uses, the majority of annual production of each CFC is emitted relatively promptly, and each has about the same depletion efficiency per unit mass (Table 1). For carbon tetrachloride, the weight is 0.064, the estimated share of production that is emitted (most carbon tetrachloride produced is consumed in the production of CFC-11 and 12). Carbon tetrachloride also has about the same depletion efficiency as have the three CFCs. The factor for methyl chloroform is 0.1, since atmospheric models suggest its effect on ozone is an order of magnitude smaller than the three CFCs. Like the CFCs, emissions are typically prompt. Finally, a factor of 10 is used for both Halons, since estimates available at the time (Quinn et al., 1986) suggested their depletion efficiency might be an order of magnitude or more greater than those of the CFCs above. More recent calculations (Hammitt et al., 1987) estimate a depletion efficiency of 11.4 for Halon 1301 but only 2.7 for Halon 1211 (Table 1).

The most important result of using this weighting scheme is that the growth rates of chemicals produced in large volume, or likely to have a larger depletion effect per unit mass, contribute more to the score function than others. As a result, the score function should serve as a proxy for the potential magnitude of ozone depletion.

From Score Function to Component Values

Any value of the score function corresponds to an infinite variety of production growth rates for individual chemicals. This problem lies at the core of using scenarios; no one seriously expects any one scenario to occur in the sense that all growth rates specified in the scenario persist as expected over the life of the scenario. Scenarios are designed to illustrate the implications of an underlying probability distribution or to represent some general kind of event that has a significant probability of occurring. A simple convention is used to pick the single set of time profiles that comprise a scenario.

The scenarios are intended to illustrate the different time profiles of ozone depletion that production of the seven PODs might induce; the specific time profiles for individual chemicals are of secondary interest. It seems reasonable to use a convention that treats sources of uncertainty equally to avoid any manipulation of the approach aimed at emphasizing one compound over another. Accordingly, the convention chosen is to use the same quantiles for each of the independent component distributions. Other conventions could be chosen, just as scenarios with similar policy implications can be defined in different ways.

CHAPTER 4
THE DISTRIBUTION OF THE SCORE FUNCTION

Implementing the approach requires explicit subjective probability distributions for production growth rates and a method for constructing the distribution of the score function. In this section, a formulation is developed that allows a closed analytic solution to derive the distribution of the score function, and to trace back to individual chemical production. The subjective probability distributions for the individual chemicals are described in Chapter 5.

The method approximates the distribution for the score function that would be developed if all the component variates that describe the growth rates of individual PODs were distributed normally.[1] Normal distributions appear to reasonably represent beliefs about future use of each POD in all but a few instances, and these can be accommodated without serious difficulty. The approximation would be exact if the relative production shares of each chemical remained constant over time. These shares change, but not enough to seriously threaten the integrity of the general results.

The general conceptual approach outlined in Chapter 3 would allow a more general implementation based on standard Monte Carlo techniques that would require neither an assumption of approximate normality nor the kinds of approximations used here. Whether this approach would justify the additional costs is unclear, since a Monte Carlo analysis requires approximations when continuous distributions are approximated by discrete ones.

AGGREGATING CHEMICAL USE

The growth rate for the production of each chemical is characterized as the sum of the growth rate of general economic activity (Gross Domestic Product or GDP) and of the intensity of use of each chemical, defined as the level of use relative to GDP.[2] The subjective probability distributions for the average annual growth rates of these components are normal distributions. Because the score function is a linear combination of the random component variables, uncertainty about its value is also described by a normal distribution. The parameters of the distribution of the score function's value can be expressed as functions of the parameters of the distributions of the component variables.

Initial distributions for use of the seven PODs before 2000 are divided by world region and, for CFC-11 and 12, by product applications. These distributions must be aggregated to derive the distributions for world use of each chemical, and similarly aggregated across chemicals to derive the distribution for the score function. Linear approximations are used in making these aggregations.

First consider the aggregation of growth rates for different uses of a chemical or for total use of a chemical in different world regions. For example, subjective probability distributions for CFC-12 use in aerosols, foam blowing, refrigeration, air conditioning, and other applications have been developed (Hammitt et al., 1986). Let x_i be the amount of CFC-12 used in the ith application. Then the total amount of CFC-12, x, is simply $\Sigma\ x_i$ and the rate of change in x can be related to the rate of change of the $\{x_i\}$ at any instant by

(1) $(1/x)(dx/dt) = \Sigma_i\ [(x_i/x)(1/x_i)(dx_i/dt)].$

Define r as the growth rate of CFC-12, r_i as the growth rate of the ith application of CFC-12, and a_i as the share of CFC-12 used in the ith application ($a_i = x_i\ /\ \Sigma\ x_i$). Then equation (1) can be reexpressed as

(2) $r = \Sigma_i\ (a_i\ r_i).$

Given the small growth rates under consideration, $r_i \approx g + u_i$, where g is the growth rate for GDP and u_i is the growth rate for the intensity of use of the ith application relative to GDP. Hence, equation (2) can be rewritten as

(3) $r \approx g + \Sigma_i \, (a_i \, u_i).$

The derivation of equation (3) incorporates two linear approximations. First, it assumes that the growth rate of a variable can be approximated by the sum of the growth rates of the two factors affecting it. Given the small growth rates under consideration, this is adequate. Second, while equation (2) is exact at any instant, it is not exact over a discrete time interval unless all u_i are equal and hence all a_i remain constant over the period. The $\{a_i\}$ corresponding to the beginning of the period are used (corresponding to a Laspeyres index of aggregated growth rates); this index is adequate so long as the $\{a_i\}$ do not shift too much over the period of interest.[3]

Similarly, a linear formula is used to aggregate use across world regions to approximate the global growth rate. If r_j is the rate of growth in the jth region and b_j is the share of global use there, the global growth rate

(4) $r \approx \Sigma_j \, (b_j \, r_j).$

The $\{b_j\}$ are fairly stable as long as uses in different regions do not grow at markedly different rates.

The second important linear approximation is in the definition of the score function. The score function is defined as

(5) $s = \Sigma_k \, (w_k \, r_k)$

$\approx g + \Sigma_k \, (w_k \, u_k)$

where w_k is the weight for the kth POD (incorporating the share of production that is emitted and the relative depletion efficiency) and u_k is the growth rate for its intensity of use.

DERIVING SCENARIOS FOR INDIVIDUAL CHEMICALS

If g and u_k in equation (5) are normally distributed, then s is normally distributed as well. This observation is the key not only to convoluting uncertainties in the components into a distribution for the score function, but also to deriving POD use rates consistent with a scenario defined by any value of the score function. Once a scenario is defined in terms of a quantile of the distribution for s, the normality of the score function and its components can be used to find the common quantile for the component distributions that is consistent with this scenario.

Begin by noting that the mean (m_s) and variance (v_s) of s can be defined in terms of the means (m_k) and variances (v_k) of its components, and the covariance among components $(v_{k\lambda})$:

$$(6) \quad \begin{cases} m_s = m_g + \Sigma_k (w_k m_k) \\ v_s = v_g + \Sigma_k (w_k^2 v_k) + 2 \Sigma_{k>\lambda} (w_k w_\lambda v_{k\lambda}). \end{cases}$$

The value of the score function at the q_sth quantile of its distribution is

$$(7) \quad s(q_s) = m_s + v_s^{0.5} z(q_s)$$

where $z(q_s)$ is the value of the z-statistic corresponding to the q_sth quantile of the standard normal distribution. Analogously, the q_cth quantiles of the growth rates of GDP and the intensity of use of the ith chemical can be defined as

$$(8) \quad \begin{cases} g(q_c) = m_g + v_g^{0.5} z(q_c) \\ u_i(q_c) = m_i + v_i^{0.5} z(q_c). \end{cases}$$

The value $z(q_c)$, and thus implicitly the quantile q_c, is determined so so that if g and all the u_k are fixed at their q_cth quantiles, s will take the value at its q_sth quantile. To do this, for a given value of s, substitute expression (8) into (5) and (6) into (7) and set (5) equal to (7). Rearranging yields

(9) $z(q_c) = z(q_s) / \beta$

where

$$\beta = \frac{v_g^{0.5} + \Sigma(w_i \, v_i^{0.5})}{[v_g + \Sigma(w_i^2 \, v_i) + 2\Sigma_{k>\lambda}(w_k \, w_\lambda \, v_{k\lambda})]^{0.5}}.$$

Equation (9) allows one to easily transform quantiles of the distribution of the score function into the corresponding quantiles of the distributions underlying the POD growth rates. Each scenario is based on a value of q_s, which yields $z(q_s)$, from which $z(q_c)$ can be calculated using Eq. (9).[4] It can be shown that $\beta \geq 1$ so that the quantiles for the component distributions are closer to their medians than the corresponding quantiles for the distribution of the score function.[5] That is, the quantiles of the component variates that correspond to a specified subjective probability interval for the score function will span an interval associated with a lower level of subjective probability. As shown in Table 6 below, the quantiles of the component distributions corresponding to the 90-percent subjective probability interval for the score function approximately span a 65-percent subjective probability interval for each component variate.

CHAPTER 5

SUBJECTIVE MARGINAL PROBABILITY DISTRIBUTIONS
FOR POTENTIAL OZONE DEPLETERS

This section explains how subjective probability distribution for the production growth rates of the seven chemicals over the period 1985 to 2040 were chosen. It begins with a brief explanation of the methodology used to develop the distribution from information reported elsewhere (Hammitt et al., 1986; Quinn, et al., 1986), then summarizes the analysis for the periods 1985-2000 and 2000-2040.

The object of interest in developing the subjective probability distribution is how the emissions of the seven PODs grow over time. To examine this, I focus on production growth rates and infer emissions from simulated production, taking account of the distribution of POD use across applications.

The structure of correlations among growth rates for the seven chemicals at a single time, and for single chemicals over time, may be complex. Both substitution and complementarities are observed among specific PODs in certain applications. For example, either CFC-11 or 12 can be used as a refrigerant in some applications, and each can be used to produce similar foam products. In household refrigerators, the compounds are complementary: CFC-12 is used as the refrigerant and CFC-11 is used in the foam insulation. Similarly, methyl chloroform and CFC-113 are substitutes in some solvent applications, but their use may be positively correlated because both are subject to similar waste-disposal regulations. Specifying these relationships completely would entail a level of detail not warranted by current knowledge of the relationships. Instead, a simplified specification that captures the most important features is adopted.

Relationships across Time

Because the PODs' atmospheric lifetimes are measured in decades, annual variation in emissions is not important: the primary concern is with cumulative emissions over long periods. Thus, long-term average annual growth rates are analyzed. Specifically, means and variances are chosen for the distributions for the secular growth rate for each chemical and covariances among growth rates for different chemicals in each period. Because of the focus on long-term average rates, address year-to-year variations associated with the business cycle or transient market conditions need not be addressed. It is reasonable to ignore these events because their influence on the variance of the average growth rate falls as the length of the period grows,[1] and should be quite modest over the 15-year and longer periods considered.

The period from 1985 to 2040 is divided between two subperiods and the means and covariance structure of secular growth within each period are specified separately. Parameter values for the period 1985-2000 are based on subjective judgments about the range of reasonable growth rates for chemicals in different applications reported in Hammitt et al. (1986). These judgments are based on detailed analysis of market trends and potential changes in markets, technologies, and regulations that could affect use of these chemicals.

Parameter values for the period 2000-2040 are based on concepts developed in Quinn et al. (1986) and comparison with the results for 1985-2000. Quinn et al. (1986) examine historical trends in the relationship between chemical use and national income and use these to project a range of use levels over the period in question. The analysis is necessarily less detailed than that in Hammitt et al. (1986). Because POD use in the distant future is less certain than in the near future, these results are adjusted to provide greater annual variance in the period after 2000.

Although the structure of growth rates within each period may be conceived as a set of means and a covariance matrix for each period, the relationship between the periods is more difficult to represent parametrically. A relationship that incorporates the implicit assumption of a positive correlation between annual growth rates within the two periods is used;[2] a "high growth" scenario for the full period includes high growth scenarios in each period. Similarly, low growth scenarios use low growth rates in both periods.

Relationships between Chemicals

The average annual growth rate is treated as the sum of two terms: (1) general economic growth (GDP) and (2) growth of intensity of use relative to GDP, which includes both growth in the product markets where the chemical is used (relative to GDP) and growth in the use of the chemical per unit of product. Sources of uncertainty in the intensity of use of each chemical are assumed to be uncorrelated between chemicals, both within and across periods, although it would be possible to incorporate relationships of this kind by introducing additional component variates.

The general economic growth component is common to all of the chemicals, and creates a positive covariance among the growth rates for all of the chemicals. Before 2000, the world is divided into regions and general economic growth and chemical intensity distributions are specified for each; after 2000, the world is treated as a single unit. When dealing with more than one region, growth rates in different regions are assumed to be positively correlated with one another. The next subsection discusses this in more detail.

In sum, the method characterizes a wide variety of factors relevant to uncertainty about the future use of these chemicals. It reflects interrelationships across time and across chemicals. A more complicated framework for relating growth rates could potentially capture subtleties not represented here. However, empirical data of the type and quality necessary even to quantify all of the details of this system in an historical period are not currently available. Until better data are available, a more detailed structure does not appear appropriate.

SUBJECTIVE PROBABILITY DISTRIBUTION FOR THE PERIOD BEFORE 2000

The subjective probability distribution for the period before 2000 is taken from Hammitt et al. (1986). That source develops explicit 80-percent subjective probability intervals for the use and production of the seven potential ozone depleters through the end of the century, taking account of uncertainty about general economic activity, demand for specific products relative to it, and POD use per unit product. It divides world use into use in three major regions, and projects future use of CFC-11 and 12 in each of the major products in which they are used. The mean rates of growth are based on analysis of trends and industry forecasts for each application or chemical. Although the projected mean rates are not constant over the period for some of the uses, average annual rates are used here.

Distributions for growth rates of regional economies, applications of single chemicals within a region, and chemical use in each region are aggregated to obtain the distributions for Gross Global Product (GGP) and chemical intensity growth using the methods described in Chapter 4. When aggregating distributions across regions, a positive correlation coefficient (0.75) is used across regions for both GDP and chemical intensity growth rates. The parameters of the resulting distributions are reported in Table 3.

The mean of the distribution for the production growth rate of each POD is the sum of the means of the distributions for general economic growth and intensity of use; the standard deviation is the square root of the sum of the variances. The covariance among the production growth rates is equal to the variance of general global economic growth (1.32) because the intensities are uncorrelated with one another and with general economic growth.[3]

As shown by Table 3, use of CFC-113 and the Halons is expected to grow more rapidly during the remainder of the century than use of the other chemicals; CFC-12 is expected to decline in use relative to the world economy, although not in absolute quantity. This distinction reflects a typical pattern of chemical use, in that relatively recently marketed

Table 3

PARAMETERS OF THE SUBJECTIVE NORMAL PROBABILITY DISTRIBUTION
FOR GROSS GLOBAL PRODUCT AND CHEMICAL USE RELATIVE TO GGP

(Average growth rates in percent per annum)

Chemical use relative to GGP	1985-2000		2000-2040	
	Mean	Standard deviation	Mean	Standard deviation
CFC-11	0.02	0.98	0	0.67
CFC-12	-1.00	1.05	0	0.72
Carbon tetrachloride	-0.49	1.02	0	0.70
CFC-113	3.27	1.67	0	1.15
Methyl chloroform	-0.32	1.10	0	0.75
Halon 1301	1.08	2.29	-0.45	1.57
Halon 1211	0.96	2.33	-0.05	1.60
Gross global product	3.28	1.15	2.40	0.96

Note: Chemical growth rates are sum of growth rates of GGP and chemical use relative to GGP. Correlations among rates are zero.

"specialty" chemicals that are produced in limited quantities grow rapidly as they are adopted in applications to which they are well suited. Older chemicals, produced in larger quantities, may not grow as quickly because they have already been adopted in the applications to which they are best suited.

SUBJECTIVE PROBABILITY DISTRIBUTION FOR THE PERIOD BEYOND 2000

The subjective probability distribution for the post-2000 period is derived from information from a number of sources. Economic growth parameters are based on data from William Nordhaus and Gary Yohe.[4] Intensity growth parameters are based on concepts and estimates developed in Quinn et al. (1986) and the parameters chosen for the period before 2000. Information for the earlier period is used to ensure consistency across periods. This subsection first considers how information from these sources was used and then reports the resulting distribution for the post-2000 period.

Development of the Subjective Probability Distribution

Reliance on detailed analyses of individual chemical applications and markets is much less appropriate in the post-2000 than the pre-2000 period. It is difficult to imagine the range of events that may occur in this period. Quinn et al. (1986) present the best available analysis of the kinds of events that may affect POD use in the next century, but even this analysis shows less proportional variation in the next century than in this. Moreover, it does not allow for variation in overall economic growth. As a result, the concepts developed in Quinn et al. (1986) serve as a foundation from which a more realistically broad range of uncertainty for this period is developed.

First, consider the distribution of GGP growth. Parameters for it are based on distributions Nordhaus and Yohe have developed for the projected rates of population and labor productivity growth. The parameters are based on a weighted average of their parameters for these distributions over the periods 2000-2025 and 2025-2050, incorporating a correlation coefficient of -0.25 between population growth and growth in labor productivity over the period 2000-2040.[5] Since the GGP growth rate is approximately the sum of the population and labor productivity growth rates, the mean GGP growth rate is the sum of the means that Nordhaus and Yohe use for population and labor productivity, and the variance is the sum of the variances that they use for population and labor productivity less a small quantity to take account of the negative correlation between them.[6]

Second, consider parameter values of the distributions for chemical use relative to GGP. Quinn et al. (1986) analyze historical POD use as a function of GDP per capita, and assume that POD use will tend to grow with per capita GDP over the long run. The estimates made here assume that POD use will maintain a constant share of the economy; that is, that it will grow at the same rate as GDP and the intensity growth rate will approximate zero. It seems more appropriate to assume that POD use in two countries of differing population but similar per capita GDP will not be equal, but will approximate the ratio of populations. This base implies that two countries of equal GDP, one poor but populous and the other rich but less populous, would use equal quantities of PODs. I believe this assumption is more plausible, since many POD applications are in products used over a wide income range.

For most chemicals, this approach implies a mean intensity growth rate of zero for the post-2000 period. The Halons are treated slightly differently. For Halon 1301, recovery and recycling from fire-extinguishing systems are expected to become an important source as the buildings that have been fitted with these systems in the 1980s and 1990s are remodeled or destroyed. Even if demand for new systems grows with GGP, intensity of use of new material should fall, although total production may continue to grow. For Halon 1211, penetration of the portable fire-extinguishing market is expected to slow as this application becomes saturated, leading to a slightly negative mean intensity growth rate. These mean intensities are reported in Table 3.

Appropriate variances are more difficult to choose for these distributions because the range of uncertainty is difficult to envision this far into the future. Variances are based on the ranges for CFC-11 and 12 suggested in Quinn et al. (1986), adjusted for compatibility with the ranges for these chemicals developed for the pre-2000 period.

Comparison of growth-rate variances across the periods is based on the concept that variation in mean growth of each chemical is due to variation in a large number of component events, many of which are related weakly, if at all. Over a short period, the number of relevant events, such as changes in regulation, technology, and product development, is likely to be small, so that a single change could have a substantial effect on the growth rate. Over a longer period, these events are likely to partially offset one another, suggesting that the range of intensity growth rates should be smaller for longer periods, assuming the number and character of changes likely to occur in any fixed time interval are comparable. This is simply a reflection of the fact that the variance of the mean of a set of identically, not-perfectly-correlated random variates falls as the number of variates increases. Because the pre-2000 period is so much shorter than the post-2000 period, it is necessary to adjust the variances for comparison.

This view of the uncertainties underlying the distribution of intensity growth is used to choose the variance of growth rates in the post-2000 period. The variance for the pre-2000 period for each chemical is multiplied by 15/40 to adjust for the difference in the length of the periods, and inflated by a factor to reflect greater uncertainty about the distant as opposed to the nearer future. This factor is somewhat arbitrary; the value 1.25 was used based in part on results in Quinn et al. (1986).

The means and standard deviations for GGP and chemical intensity are reported in Table 3. Note that the means of the subjective probability distribution for growth rates are much lower than for the period before 2000, and the standard deviations are also smaller, as discussed above. The means for the intensity growth rates for all of the PODs except Halon 1301 are essentially zero. The Halon 1301 rate is negative because of expected recovery and reuse of the compound from dismantled systems. Standard deviations are also similar for all of the PODs but CFC-113 and especially the Halons. CFC-113 use is more uncertain than others because of the possibility of regulations affecting disposal of waste solvents and substantial substitutability with other solvents. Because they are relatively new and of limited but expanding application, significantly greater uncertainties attend the future of the Halons than the future of the other chemicals.

MARGINAL DISTRIBUTIONS FOR INDIVIDUAL CHEMICAL PRODUCTION

Table 4 presents illustrative quantiles of the marginal distributions for production of individual chemicals in 2000, 2020, and 2040, together with estimated current world use. These results suggest that a wide range of future outcomes are possible. However, uncertainty about the joint effect of these compounds on calculated ozone depletion is smaller, because it is highly unlikely that use of all of the chemicals will grow at an extremely high (or low) rate. In the next section, scenarios that recognize this joint variation are derived.

Table 4

PRODUCTION LEVELS AT SPECIFIED QUANTILES OF THE SUBJECTIVE MARGINAL PROBABILITY DISTRIBUTIONS FOR INDIVIDUAL CHEMICALS

(Thousand metric tons)

Chemical/ Year	Quantile of Marginal Probability Distributions				
	.05	.25	.50	.75	.95
CFC-11					
1985	342	342	342	342	342
2000	386	479	556	644	794
2040	466	907	1435	2263	4326
CFC-12					
1985	444	444	444	444	444
2000	425	532	622	725	901
2040	504	1001	1606	2565	4994
Carbon tetrachloride					
1985	1029	1029	1029	1029	1029
2000	1070	1335	1554	1807	2237
2040	1280	2519	4014	6371	12290
CFC-113					
1985	163	163	163	163	163
2000	262	348	422	512	671
2040	256	605	1091	1956	4476
Methyl chloroform					
1985	545	545	545	545	545
2000	574	721	844	986	1229
2040	672	1349	2179	3506	6893
Halon 1301					
1985	11	11	11	11	11
2000	11	16	20	26	37
2040	7	21	44	92	259
Halon 1211					
1985	11	11	11	11	11
2000	11	16	20	26	37
2040	8	25	53	111	315

CHAPTER 6
PRODUCTION AND EMISSION SCENARIOS

Applying the methods described in Chapter 4 to the subjective probability distribution defined in Chapter 5 yields scenarios that can be used to project alternative futures for the seven PODs analyzed here. This section explains how the final calculations are made, presents a set of production scenarios based on the technique, and reports the corresponding ozone depletion calculated for these scenarios using an atmospheric simulation model.

CHOOSING QUANTILES FOR SCENARIOS

A production scenario is based on a particular quantile of the distribution of the score function. The 5th, 25th, 50th, 75th, and 95th percentiles are used as a basis for scenarios that represent respectively lower limit, low, middle, high, and upper limit cases relevant to policy decisions. Scenarios based on other quantiles could obviously be developed without difficulty using the same techniques. These five appear to describe the relevant policy space in a way that facilitates analysis.

Viewed in the context of the overall policy analysis, the middle three scenarios—low, middle, and high—may be the most useful. The middle case represents a scenario defined such that the effect of these seven PODs on ozone depletion is equally likely to be greater or smaller than the effect corresponding to this scenario. Analogously, the low and high scenarios are defined such that the probabilities of greater or lesser effects due to these chemicals, conditional on the effect being greater or smaller than the median case, are equal. Thus, these three scenarios divide the range of ozone depletion into four equal-probability intervals. The limiting scenarios at the 5th and 95th percentiles can be used to characterize extreme but conceivable outcomes.

Development of the scenarios starts with chemical weights for the score function that reflect relative production levels in 1985. Table 5 presents these weights. Together with the values of the standard deviations from Chapter 5 and equation (9) from Chapter 4, these weights allow calculation of the value β with which to transform z-statistics from the distribution for the score function into z-statistics for the component economic-growth and intensity distributions. These identify intensity and general economic growth rates that can be used to calculate a growth rate for each chemical under each scenario during the pre-2000 period. These growth rates, applied to actual production levels in 1985, are used to calculate the production paths relevant to each chemical for each scenario up to 2000.

The calculations for the post-2000 period are analogous. Different weights and consequently a different value of β are used, based on production levels in 2000 for the 50th percentile growth scenario. These are also reported in Table 5.

Median growth rates differ enough across chemicals during the pre-2000 period to lead to modest shifts in the chemical weights between 1985 and 2000. This means that the weighting system is only approximate over this period. Median growth rates after 2000 are similar for all chemicals but the Halons. Although a constant set of weights are used over a significantly longer period of time after 2000, the actual weights do not shift as much over this period as they do from 1985 to 2000.[1]

Table 5

SCORE-FUNCTION WEIGHTS

Chemical	Pre-2000	Post-2000	Approximate Difference
CFC-11	.259	.240	-0.02
CFC-12	.368	.301	-0.07
Carbon tetrachloride	.054	.047	-0.01
CFC-113	.110	.138	0.03
Methyl chloroform	.048	.041	-0.01
Halon 1301	.080	.116	0.04
Halon 1211	.080	.116	0.04

Note: The weights reflect the production level, share of production that is emitted, and relative depletion efficiency of each compound. Of these factors, only production levels differ between periods.

Applying these weights in equation (9) yields beta values of 1.69 before 2000 and 1.79 afterward. These yield the z-statistics for individual scenarios shown in Table 6. The results in this table make it clear how important it is to view chemicals jointly rather than individually. Scenarios for the fifth percentile of the score function use z-statistics for component distributions that are smaller by a factor of β (about 1.7) and are consistent with the 17th to 18th percentiles of these distributions. A 25th percentile scenario uses z-statistics consistent with the 34th to 35th percentiles of the component distributions. Similar adjustments apply for higher percentile scenarios. Viewing these seven chemicals together significantly narrows the range of growth rates represented in the scenarios before and after 2000; the effect is slightly larger after 2000.

CHEMICAL USE AND EMISSION SCENARIOS

Taken together with the parameter values from Chapter 5, the z-statistics in Table 6 yield the growth rates in Table 7 as the bases for production scenarios. Table 8 shows the production levels that result from these growth rates for 1985, 2000, and 2040.

The timing of POD emissions varies by application. In some, PODs are released to the atmosphere shortly after manufacture. In others, they remain "banked" in products until disposal. To derive emission scenarios from the production scenarios, appropriate lags between production and emission for each POD are introduced. Differences of a few years

Table 6

Z-STATISTICS AND BETAS FOR DISTRIBUTIONS
OF THE SCORE FUNCTION AND COMPONENTS

Quantile of the Score Function	z-statistics		
	Score Function	Pre-2000 Components	Post-2000 Components
0.05	-1.645	-0.973	-0.919
0.25	-0.675	-0.399	-0.377
0.50	0.0	0.0	0.0
0.75	0.675	0.399	0.377
0.95	1.645	0.973	0.919
Beta		1.69	1.79

Table 7

PRODUCTION GROWTH RATES FOR JOINT DISTRIBUTION SCENARIOS

(Percent per annum)

Chemical	Period	.05	Quantile of the Score Function .25	.50	.75	.95
CFC-11	Pre-2000	1.47	2.54	3.29	4.03	5.12
	Post-2000	0.93	1.80	2.40	3.00	3.86
CFC-12	Pre-2000	0.39	1.50	2.27	3.04	4.15
	Post-2000	0.89	1.78	2.40	3.02	3.91
Carbon tetrachloride	Pre-2000	0.93	2.03	2.79	3.55	4.64
	Post-2000	0.91	1.79	2.40	3.01	3.89
CFC-113	Pre-2000	4.15	5.56	6.55	7.54	8.97
	Post-2000	0.51	1.63	2.40	3.18	4.29
Methyl chloroform	Pre-2000	1.03	2.17	2.96	3.74	4.88
	Post-2000	0.86	1.77	2.40	3.03	3.94
Halon 1301	Pre-2000	1.12	2.94	4.07	5.34	7.15
	Post-2000	-0.20	1.07	2.00	2.92	4.22
Halon 1211	Pre-2000	1.12	2.94	4.07	5.34	7.15
	Post-2000	0.19	1.51	2.47	3.39	4.72

in the timing of emissions have little effect on stratospheric ozone depletion, since the PODs' atmospheric lifetimes are so long. However, changes in recovery practices or other conditions that affect the proportion of the POD used in a product that is eventually released will affect ozone depletion. Uncertainty about emissions, conditional on production, could be incorporated in this analysis by introducing additional component variates to account for these possibilities. For the present, however, these uncertainties are ignored and emission trajectories are calculated as a deterministic function of production. The resulting emissions are reported in Table 8.

CFC-11 and 12 applications can be divided into five emission categories: (1) Hermetically sealed applications (home refrigerators and freezers) have estimated typical lifetimes of 17 years during which small amounts of CFC-12 are emitted as individual units fail. (2) Non-hermetically sealed applications (retail-food refrigerators, building chillers,

Table 8

PRODUCTION AND (EMISSIONS) FOR JOINT DISTRIBUTION SCENARIOS

(Thousand metric tons)

Chemical/ Year	Quantile of the Score Function									
	.05		.25		.50		.75		.95	
CFC-11										
1985	342	(238)	342	(238)	342	(238)	342	(238)	342	(238)
2000	426	(309)	499	(358)	556	(396)	619	(438)	723	(507)
2040	618	(525)	1017	(819)	1435	(1122)	2022	(1543)	3294	(2445)
CFC-12										
1985	444	(412)	444	(412)	444	(412)	444	(412)	444	(412)
2000	471	(467)	555	(541)	622	(599)	696	(662)	817	(765)
2040	672	(662)	1125	(1092)	1606	(1544)	2287	(2178)	3787	(3559)
Carbon tetrachloride										
1985	1029	(66)	1029	(66)	1029	(66)	1029	(66)	1029	(66)
2000	1183	(76)	1391	(90)	1554	(100)	1736	(112)	2032	(131)
2040	1702	(110)	2827	(182)	4014	(259)	5686	(366)	9339	(601)
CFC-113										
1985	163	(138)	163	(138)	163	(138)	163	(138)	163	(138)
2000	300	(253)	367	(310)	423	(357)	485	(410)	591	(499)
2040	367	(310)	700	(591)	1091	(922)	1695	(1432)	3172	(2680)
Methyl chloroform										
1985	545	(474)	545	(474)	545	(474)	545	(474)	545	(474)
2000	636	(554)	752	(654)	844	(734)	946	(823)	1113	(969)
2040	898	(781)	1517	(1320)	2179	(1896)	3123	(2717)	5215	(4537)
Halon 1301										
1985	10.7	(3.0)	10.7	(3.0)	10.7	(3.0)	10.7	(3.0)	10.7	(3.0)
2000	13.6	(5.6)	16.7	(6.5)	19.4	(7.1)	22.4	(7.9)	27.7	(9.2)
2040	9.9	(8.5)	24.2	(14.7)	43.2	(22.0)	75.4	(33.6)	163	(62.9)
Halon 1211										
1985	10.7	(3.0)	10.7	(3.0)	10.7	(3.0)	10.7	(3.0)	10.7	(3.0)
2000	13.6	(5.6)	16.7	(6.5)	19.4	(7.1)	22.4	(7.9)	27.7	(9.2)
2040	17.5	(16.1)	34.1	(24.6)	55.1	(33.9)	89.6	(47.9)	182	(81.3)

mobile air conditioners) lose CFCs continually through leakage and servicing throughout their estimated seven-year lives. (3) Urethane closed-cell foams emit CFC-11 very slowly. During their estimated 30-year service life, emissions are assumed to decline geometrically and are characterized by an assumed 100-year half life (based on measurements by Khalil and Rasmussen, 1986). After disposal, emissions are assumed to accelerate to reflect the breakup of the foam into smaller pieces. Additional small amounts are emitted at manufacture. (4) Non-urethane foams release CFC-12 over an estimated two-year period. (5) Prompt emitters (aerosol propellants, flexible-foam blowing, sterilants) emit CFCs as they are used, shortly after manufacture.[2]

Approximately 94 percent of carbon tetrachloride produced is used as an intermediate in the manufacture of CFCs. Most of the remainder is emitted promptly.

Most CFC-113 and methyl chloroform emissions are prompt, although an estimated 15 percent of production is lost in waste-disposal dumps, incinerated, or used as intermediates in manufacturing other chemicals.

Halons are largely banked in fire-extinguishing systems that have estimated service lives of up to 40 years. Small amounts are emitted annually through leakage, fire, and inadvertent discharge, and at disposal (Hammitt et al., 1987; Palmer et al., 1980).

MODEL-CALCULATED OZONE DEPLETION

The projected ozone depletion corresponding to these joint emission scenarios can be calculated using computer-based atmospheric models. I present results using the Lawrence Livermore National Laboratory (LLNL) one-dimensional model of transport, photochemical kinetics, and radiative processes in the troposphere and stratosphere. The LLNL model is representative of the models used for calculating future ozone changes and is described by Wuebbles (1983), Connell and Wuebbles (1986), and World Meteorological Organization (1986). Its domain extends from the earth's surface to 55 km altitude and it includes 165 photolytic and thermal reactions of 52 chemical species, a diffusive representation of globally averaged transport, and interactive calculation of the temperature profile above 14 km (assuming radiative equilibrium). All significant known homogeneous photochemistry of the oxygen, nitrogen oxide, chlorine, bromine, and methane families is included, but heterogeneous reactions of the type that are believed to contribute to the severe depletion in the Antarctic springtime are not included. Kinetic rate constants and photochemical parameters are based on the recommendations of the National Aeronautics and Space

Administration (NASA) Panel for Data Evaluation (DeMore, 1985). From an initial condition representing the current (1985) atmosphere, integration over time produces the model-predicted global annual average changes in tropospheric and stratospheric ozone abundances.

The calculated change in total column ozone represents an annual global average of the seasonally and latitudinally varying changes that are expected to result from POD and other trace-gas emissions. One-dimensional model results are useful for delineating the ranges of calculated ozone change that arise from uncertainties about future POD emissions and other factors, although higher dimensional models are preferred for evaluating possible effects on the biosphere, where seasonal and latitudinal differences in UV flux may be important.

To calculate the predicted effects on ozone of the joint POD emission scenarios, trends for other atmospheric trace species must be assumed. The ambient 1985 atmosphere is obtained using industrial emission estimates for the CFCs and observed trends for CO_2, N_2O, and CH_4 abundances as boundary conditions (Connell and Wuebbles, 1986). Trends in trace-gas abundances include the carbon dioxide projection in Edmonds et al. (1984) (averaging 0.64 percent/year) and assumed continuation of currently estimated global average trends in nitrous oxide (0.25 percent/year) and methane (1 percent/year) (World Meteorological Organization, 1986). Surface emissions of HCFC-22 are fixed at the current rate of about 71.9 thousand metric tons/year. The effects of uncertainties in these assumptions on calculated ozone change are discussed by Connell and Wuebbles (1986).

Changes in model-calculated total column ozone for the joint emission scenarios are shown in Fig. 4, and the corresponding calculated surface abundance of the seven PODs and other trace gases at selected dates are reported in Table 9. The figure also shows the calculated changes if POD emissions remain constant at 1985 rates, and if one proposed policy, a cap on CFC-11 and 12 production at the estimated current world capacity (1.24 million metric tons; Mooz et al., 1986) is adopted and satisfied. The effect of this policy would depend on how rapidly POD use grows, and thus on how quickly the cap is reached (and on how rapidly production and emissions of the other PODs continue to grow). The case illustrated assumes production follows the 50th-percentile scenario until the cap is reached in 2003, but a more complete analysis of this policy would consider the distribution of effects that would result, corresponding to POD use growing in accordance with other scenarios as well.

The range of calculated ozone loss corresponding to the 5th- and 95th-percentile scenarios spans an order of magnitude (1.8 percent to 17.3 percent) by 2040. At 6.2 percent, the 50th percentile scenario falls nearer the geometric than the arithmetic mean of the other scenarios, as the distribution of ozone depletion is skewed to the smaller depletion side. This skewness reflects the skewness of emission levels, resulting from the symmetry of growth rates across scenarios. The 25th- and 75th-percentile scenarios correspond to depletion of 3.9 and 9.4 percent.

Even if POD emissions are held to 1985 levels, POD abundances continue to increase because of the long atmospheric lifetimes. The initial small decrease in column ozone abundance is followed by a continuing small increase as carbon dioxide and methane abundances increase. Increasing carbon dioxide cools the stratosphere and slows ozone-loss processes. Increasing methane has the net effect of increasing ozone as a by-product of methane oxidation in the lower stratosphere and troposphere, and by sequestering active chlorine species in the inactive form of hydrochloric acid (HCl).

Capping CFC-11 and 12 production at estimated current capacity in the 50th-percentile scenario allows production of these species to increase until 2003, after which they remain constant. Ozone depletion continues, reaching 4.9 percent in 2040, about 1.3 percentage points less than under the uncapped scenario. Stratospheric inorganic chlorine abundances continue to increase because of increasing emissions from the bank, the CFCs' long atmospheric lifetimes, and increasing emissions of other PODs.

Because of the method used to develop the emission scenarios, the range of calculated ozone depletion represents a subjective, approximate 90 percent probability interval for the calculated joint effect of the seven PODs on stratospheric ozone, assuming no additional regulations. It represents that part of the uncertainty about future depletion associated with current uncertainty about future uncontrolled POD emissions. A probability interval for future ozone depletion requires assessment of other factors that contribute to that uncertainty, including trends in abundances of other gases, the photochemical and kinetic processes modeled, and possible regulations.

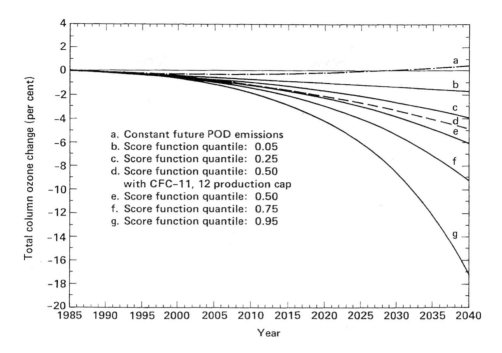

Fig. 4 — Calculated one-dimensional total column ozone change for seven POD emission scenarios. Assumed growth rates are consistent with current trends: CO_2, 0.64%/yr; N_2O, 0.25%/yr; CH_4, 1%/yr; CFC-22, 0%/yr.
Source: Hammitt, et al. (1987).

Table 9

CALCULATED SURFACE ABUNDANCES FOR JOINT EMISSION SCENARIOS

(Mole fraction)

Chemical	Year	Zero Emission Growth	Quantile of the Score Function					.50 with Production Cap
			.05	.25	.50	.75	.95	
			(Parts per billion)					
CFC-11	1985	0.206	0.206	0.206	0.206	0.206	0.206	0.206
	2000	0.300	0.332	0.346	0.356	0.367	0.385	0.356
	2040	0.508	0.770	0.995	1.21	1.48	2.01	0.860
CFC-12	1985	0.377	0.377	0.377	0.377	0.377	0.377	0.377
	2000	0.626	0.653	0.676	0.694	0.713	0.742	0.694
	2040	1.18	1.46	1.87	2.27	2.77	3.77	1.66
Carbon tetra-chloride	1985	0.153	0.153	0.153	0.153	0.153	0.153	0.153
	2000	0.158	0.161	0.164	0.167	0.169	0.174	0.167
	2040	0.166	0.199	0.249	0.297	0.358	0.480	0.297
CFC-113	1985	0.033	0.033	0.033	0.033	0.033	0.033	0.033
	2000	0.089	0.113	0.124	0.132	0.141	0.156	0.132
	2040	0.199	0.364	0.538	0.718	0.967	1.51	0.718
Methyl chloroform	1985	0.138	0.138	0.138	0.138	0.138	0.138	0.138
	2000	0.175	0.191	0.211	0.226	0.242	0.268	0.226
	2040	0.201	0.303	0.478	0.637	0.861	1.31	0.641
			(Parts per trillion)					
Halon 1301	1985	0.00	0.00	0.00	0.00	0.00	0.00	0.00
	2000	3.39	4.99	5.43	5.75	6.13	6.73	5.75
	2040	10.3	22.8	31.8	40.6	52.9	80.6	40.6
Halon 1211	1985	0.00	0.00	0.00	0.00	0.00	0.00	0.00
	2000	2.10	3.23	3.55	3.78	4.06	4.50	3.78
	2040	3.15	14.9	19.8	24.8	31.8	46.8	24.9

	Other trace gases (all scenarios, parts per million)		
	CO_2	CH_4	N_2O
1985	345	1.74	0.303
2000	372	2.03	0.315
2040	490	3.02	0.348

Source: Hammitt et al. (1987).

CHAPTER 7
CONCLUSIONS

The significant possibility that emissions of certain commercially important chemicals may deplete stratospheric ozone poses a difficult policy problem. To formulate constructive policy, it is useful to understand the many uncertainties that affect the relationship between decisions to reduce the global use of potential ozone depleters and the effects that such reduction might have on human health, materials degradation, crop yields, and other activities of interest. Only a small subset of these uncertainties are addressed here. But the relationship of these uncertainties to other parts of the policy problem provides the basic motivation. On the one hand, the complexity of the problem is so great that it must be broken into pieces if one hopes to produce useful results. On the other, the characterization of uncertainties in any part of the problem is likely to be more useful if it properly reflects the concerns of the problem as a whole. This method can be used to build scenarios relevant to one piece of the problem that relates them back to the problem as a whole.

This section addresses uncertainties associated with the production of seven PODs. Ideally, one would like to develop a subjective probability distribution for these chemicals and convolute the uncertainties reflected in this distribution with other sources of uncertainty relevant to stratospheric ozone depletion. In such an approach, developing a probability distribution for the PODs would be one step in a process to develop a distribution for ozone depletion itself.

Calculation costs dictate that only a limited number of cases be considered in the atmospheric models used to study ozone depletion. Hence, the cases considered should be chosen to embody as much information as possible. Since the distribution of ozone depletion implied by current understanding of POD use and other factors cannot be investigated directly, the cases used to examine parts of the distribution should contain information about where they lie in the distribution and about the probability density of the distribution in their vicinity. The method presented provides a way to relate scenarios for the future production of seven PODs to the probability distribution for ozone depletion.

The specific method is simple; the concept on which it is based could be used to develop more complex methods that might be more satisfactory. It remains for future analysis to determine how much improvement additional complexity would allow. For now, the specific method developed can be thought of as both an illustration of a more general conceptual approach and a practical way to implement that approach until a better method is developed.

I summarize with a brief overview of the approach: Characterize uncertainty about the growth rates of general economic activity and intensity of POD use relative to it with independent normal probability distributions. Specify values for the means and variances of these distributions. Define a policy-relevant score function as a linear combination of these growth rates. Calculate the mean and variance of its subjective probability distribution. Define scenarios in terms of quantiles of the distribution of the score function. Identify growth rates in the component distributions that are compatible with the value of the score function for the quantile defining each scenario, using a simple convention. Use the growth rates of the component distributions for a scenario to calculate the growth rates for each chemical in that scenario.

The key to this approach is the score function. It provides a policy relevant scale with which to compare alternative scenarios for seven chemicals along a single dimension. The score function should be related to the ozone depletion likely to result from the scenarios it is used to describe. The relationship may be crude, but the scale reflects the relative danger associated with different chemicals and hence the joint danger associated with any set of production levels of these chemicals. More complicated score functions could be considered if a Monte Carlo technique were used to convolute uncertainties about production rates. Ironically, as the score function comes closer to approximating the actual joint effect of a set of chemicals on ozone depletion, using it to develop scenarios may become less attractive. That is because a direct approach to the subjective probability distribution for ozone depletion becomes more attractive, eliminating the need for developing scenarios.

The specific approach relies on a number of analytic simplifications, in particular restricting the subjective distributions to normal distributions and using linear approximations. These restrictions simplify the analysis considerably, and make possible an analytic solution to the problem of convoluting the uncertainties. Abandoning normality or some other parametric distribution would provide greater freedom to express uncertainties, but would require the use of a simulation technique like Monte Carlo to convolute them. Once this step is taken, linear approximations are probably no longer needed. Simulations themselves typically require simplifying assumptions and approximations to implement them at a reasonable cost. Whether the approximations associated with normal distributions and linear aggregations here induce more serious errors than the simplifications and approximations required by a Monte Carlo simulation is an empirical question deserving closer attention.

Whether an analytical or Monte Carlo approach is taken, the interrelationships among chemicals deserve more attention. For example, CFC-11 and 12 are produced together. While the proportions in which they are produced are variable, it would be surprising if cost considerations did not induce a positive correlation in their production rates. CFC-113 and methyl chloroform are substitutes, but both are subject to similar government regulations. Changes in markets and regulations that underlie the scenarios used here could induce either a negative or positive correlation in their growth rates. These and other considerations suggest that future efforts to build scenarios for these chemicals should give closer attention to the relationships in intensity of use for different chemicals, a step this technique will accommodate.

This technique can produce an infinite set of scenarios. The ones presented are based on simple quantiles of the score function. The scenarios associated with the 25th, 50th, and 75th percentiles of its distribution characterize the relevant policy space and can represent a reasonable range of effects of market and technological developments in the context of policy analysis.

Developing scenarios for other factors conditional on the rate of general economic growth might also be worth exploring. That is because the level of the effects of ozone depletion on human activities like materials degradation and crop yields is likely to depend on the general rate of economic growth: Higher chemical growth rates will presumably have larger effects on ozone depletion which in turn will have more effect on crop yields if high economic growth has created a demand for more crops. Such relationships may prove to be quite important in sorting out the joint effects of different sources of uncertainty. The technique could accommodate scenarios conditioned on general economic growth by calculating the moments for score functions conditional on economic growth rates; how those growth rates would be chosen remains a problem.

In the end, this approach offers a simple solution to a complex problem. More complex solutions may well justify their additional costs but that is not immediately clear. Additional attention to substantive issues associated with the construction of scenarios for application to the issue of potential ozone depletion is likely to be more productive in the short run.

B. DECISION MAKING UNDER DIMINISHING UNCERTAINTY

CHAPTER 8
TIMING RESPONSES TO POTENTIAL STRATOSPHERIC OZONE DEPLETION

The "optimal" level of POD emissions that will maximize social welfare by balancing the costs of emission control against the benefits of decreased environmental damage cannot be determined, because of pervasive uncertainty about the likely extent of future ozone depletion, its relationship to the quantity of potential ozone depleters emitted, the effects of ozone depletion in the biosphere, and the appropriate valuation of these consequences. Moreover, we can expect to learn more about each of these areas of uncertainty through continuing scientific research, and through observation of atmospheric and biospheric responses to past and current POD emissions.

Because we can expect these uncertainties to diminish over time, the appropriate policy is likely to be an incremental one: The essential question is whether to add to existing regulations (the 1970s aerosol bans in the United States and several other countries) to further restrict emissions now or to make further restrictions contingent on future scientific developments.[1] The risk to delay is that, if significant emission reductions become necessary to prevent serious adverse consequences, their cost may be much larger than if emission reductions begin sooner. In the worst case, delaying further emission restrictions may unavoidably commit the earth to significant ozone depletion and climate modification. On the other side, the risk to adopting further restrictions now is that these restrictions may later prove to have been unnecessary; the costs incurred would have produced no benefits. The question is analogous to that of whether to purchase insurance: By imposing additional regulations now, immediate costs are incurred in exchange for a potential reduction in the costs of preventing and adapting to future ozone depletion.[2]

This section attempts to provide some insight to this question. I develop a general formulation of the policy question, which can be conceived as an infinite horizon stochastic dynamic program with learning. This formulation clarifies the issues, but is mathematically intractable. To provide more explicit guidance, I take advantage of specific features of this problem to develop a simplified decision framework. Because of the long time constants in the relationships between POD production and emissions, ozone depletion, and effects in the biosphere, the policy choice can be structured so the environmental damages and benefits are approximately the same under each policy. Because the benefits can be made equal, the framework focuses attention on a comparison of the expected economic costs of alternative regulatory strategies.

The simplified framework characterizes how the expected economic cost associated with alternative strategies—adopting interim regulations now, or awaiting future scientific progress before deciding whether to restrict emissions—depend on the probability that it will be worthwhile to restrict emissions. Specifically, the degree of belief about the severity of the threat of ozone depletion required to make the interim-regulation strategy cost-effective is identified. This required "degree of belief" can function like a standard of proof in judicial proceedings: It defines the confidence policy makers should have in the proposition that, without further emission restrictions, ozone depletion will occur and produce substantial adverse effects, if they are to favor additional, near-term regulations. In this sense, the decision framework helps to clarify the current policy choice.

Because the alternative regulatory strategies considered in this framework are constructed to produce equivalent ozone depletion, it is not necessary to estimate the relationship between ozone depletion and damages in the biosphere, nor to value these damages. This feature considerably reduces the information required to reach a decision and simplifies the analysis, since the economic costs of alternative control strategies may be compared directly. An important limitation of the analysis, however, is that it ignores the possibility that we may not learn whether ozone depletion is likely to produce severe consequences until it is too late to prevent it.

This section addresses the question of whether the world as a whole should impose additional regulations on emissions of potential ozone depleters now or await future developments before deciding. It abstracts from the important issues associated with coordinating action among nations. Because the costs of preventing ozone depletion, and the consequences if it does occur, are likely to vary widely among nations, these issues may be important in determining how to structure regulations and how much ozone depletion and climate modification to risk. However, the question of whether to adopt additional regulations now is logically prior to that of how to structure such rules.

CHAPTER 9
THE DECISION FRAMEWORK

This chapter formalizes the policy problem as a stochastic dynamic program with learning. This formulation aids understanding of the problem, but it is mathematically intractable. Subsequently, I present a simplified decision framework that is tractable and provides useful guidance.

THE GENERAL PROBLEM

The policy problem of whether and when to restrict POD emissions can be formulated as a stochastic dynamic program with learning over an indefinite, perhaps infinite, horizon. Define the following notation:

t, discrete measure of time (e.g., years). t increases with calendar time.

s_t, the state variable, describes the atmospheric burden of PODs, chlorine, and bromine at time t. s_t is a complex function of historical emissions of PODs and other trace gases (e.g., N_2O, CH_4, CO_2) from anthropogenic and other sources, past solar activity, and perhaps other factors. It may be adequate to approximate it by $s_t = \Sigma_\lambda \ a_s(\lambda) \ e_{t-\lambda}$, that is, as a weighted sum of past POD emissions, where the weights depend on the PODs' atmospheric lifetimes, relative depletion efficiencies, and possibly on other factors.

x, the state of nature, characterizes the relationship between PODs and biospheric damages. x incorporates the relationships between PODs and ozone depletion, between ozone depletion and increased UV flux, and between increased UV flux and damages in the biosphere.

$D(s_t, y_{dt}, x)$, current damage from ozone depletion. D depends on the state s_t, selected aspects of current technology y_{dt} (e.g., mitigating technologies like improved sunscreens, UV-resistant crops and coatings), and the state of nature x.

p_t, current POD production, assumed equal to current use.

e_t, current POD emissions. $e_t = \Sigma_\lambda \, a_e(\lambda | y_{et}) \, p_{t-\lambda}$, that is, e_t is a weighted sum of past production where the weights describe the time path of emissions from each POD application. These weights depend on aspects of current technology y_{et} that affect the timing of emissions from specific applications and the share of PODs produced that are eventually emitted.

$B(p_t | y_{bt})$, economic benefits from current POD use p_t. The benefits may depend on certain aspects of current technology y_{bt}. Technological innovation may increase or decrease $B(p_t)$ by finding new uses or substitutes for PODs.

$f_t(x | y_{xt})$, subjective probability distribution function for x. It depends on current knowledge about the relationships between PODs, ozone depletion, UV flux, and effects on the biosphere y_{xt}. It is assumed that scientists and policy makers can agree on a common probability function to characterize current understanding of these relationships. $f_t(x | y_{xt})$ is revised in each period in accordance with Bayes' rule, reflecting any knowledge gained during the period.[1] In general, the rate at which knowledge increases will be a stochastic function of cumulative research expenditures. It may also depend on past and current POD emissions, since if emissions are greater, any effects on the atmosphere will be larger, detectable effects will occur sooner, and so knowledge about them will be developed more rapidly. In the opposite case, if POD emissions are eliminated we may never learn whether significant ozone depletion would have occurred.

ϕ_t, current investment in technology and knowledge. ϕ_t is the sum of current expenditures on research addressing each area of technology or knowledge, ϕ_{xt}, ϕ_{bt}, ϕ_{et}, and ϕ_{dt} (research or development related to the effects of POD emissions, beneficial uses, emission controls, and adaptive technologies to limit damages from ozone depletion, respectively). Knowledge and technology in specific areas y_{xt}, y_{bt}, y_{et}, y_{dt} are all cumulative functions of past investments, for example, $y_{bt} = \Sigma_\lambda \phi_{bt-\lambda}$. (This formulation assumes no decay of knowledge and technology, although this could be readily incorporated.) Relationships describing the productivity of research and development expenditures ϕ on D, B, e, and f(x) are implicit in the functions D, B, e, and the updating rule for f(x).

State transition function:

$$s_{t+1} = s_t + a_s(0)e_t - \Sigma_\lambda [a_s(\lambda) - a_s(\lambda+1)] e_{t-\lambda}.$$

If s_t can be represented as a weighted sum of past POD emissions, the new state depends on decomposition of the old atmospheric burden plus current POD emissions, as specified.

Value function:

$$V_t(s_t,y_t,p_t) = B(p_t,y_{bt}) - E|f_t [D(s_t,y_{dt},x)] - \phi_t + \delta E|f_t [V^*_{t+1}(s_{t+1},y_{t+1})]$$

where $V^*_t(s_t,y_t) = Max_p [V_t(s_t,y_t,p_t)]$ and δ is the one-period discount factor. (The value function is also known as the fundamental recursive relation or the Bellman equation.)

The policy problem is to maximize $V_t(s_t,y_t,p_t)$ in the current period. The decision variables are current POD production and use (p_t) and current expenditures on research and development (ϕ_t), allocated among projects to refine estimates of the causes and consequences of ozone depletion (ϕ_{xt}), damage mitigation strategies (ϕ_{dt}), POD substitutes (ϕ_{bt}), and emission reducing measures (ϕ_{et}). Not all research and development expenditures will be funded out of public resources, but to account for changes in social welfare the real social cost of these projects must be included, regardless of the direct funding agency.

At each state, current damage $D(s_t,y_{dt},x)$ is observed. Over the near future, this observable damage is expected to be zero, or nearly zero. At present, it is not clear whether any ozone depletion that may have occurred can be distinguished from natural fluctuations. Ozone depletion may have occurred, and undetected increases in UV flux may already be producing cumulative damage, e.g., increasing future skin-cancer incidence. However, D is defined as including only currently observable damage (e.g., current skin cancers). The cumulative effects of UV flux on skin cancers are reflected in the dependence of D on s_t, that is, on cumulative emissions, and thereby on past ozone concentrations.

Observation of D provides information about x, leading to revision of $f_t(x|y_{xt})$, but uncertainty about x remains. In addition, f_t may be refined by research on the photochemical mechanisms of ozone depletion and the physical, chemical, and biological mechanisms through which ozone depletion increases UV flux which in turn damages life forms and valuable materials. Thus, at each stage t the subjective distribution function f_t shifts.

The horizon is indefinite. It may be infinite, or effectively so, if the possibility of ozone depletion represents a permanent constraint on human activities. In this case, in the very long run, optimal POD emissions may be equal to some constant, non-zero equilibrium level at which stratospheric ozone concentrations remain constant. This POD carrying capacity of the geosphere may be dependent on the chosen equilibrium stratospheric ozone concentration. At present, it is not clear whether the stratospheric ozone concentration is dynamically stable above certain POD emission levels. If the long-run solution is a constant non-zero POD emission level, the question of how quickly this equilibrium is approached is a policy question that involves balancing the costs of emission reductions against those of transient and possible permanent environmental damage.

The formulation of the problem suggests that one might change the policy affecting POD use and emissions in every period, tightening and relaxing emission controls as appropriate. A number of factors limit the desirability of this type of fine optimization, however, and complicate solution of the dynamic program. For example, the benefits and damages associated with POD use and the atmospheric response to POD emissions may exhibit discontinuities, non-convexities, indivisibilities, and irreversibilities. If a policy to reduce POD emissions induces all firms to withdraw from an industry, subsequent relaxation of the policy may not lead any to re-enter. If all the physical and human capital sunk in an industry is not salvageable, the industry may be profitable before the policy, but not sufficiently profitable to induce investors to re-invest in these non-salvageable components when the emissions policy is relaxed. Analogously, requiring firms to install modest emission-limiting equipment may be inefficient if there is a substantial chance that more effective equipment may be necessary later, assuming the first set of equipment cannot be salvaged or upgraded if necessary. In this case, the costs sunk in the first set of equipment would be lost when the second set is required.[2]

The perception that regulations will be subject to frequent revision could also provide a disincentive to invest in physical or human capital that is not fully salvageable, since investors have little assurance of being able to recover their costs before new emission restrictions limit their output or increase production costs. As a result, existing industries may fail to invest in new, more efficient technologies and consequently operate above their long-run cost curves. Similarly, potentially profitable, social-welfare increasing industries may not be established.

In the atmosphere, ozone may be slow to restore after it is depleted. Although moderate ozone depletion is likely to be reversible if POD emissions are severely restricted, if the atmosphere is perturbed too far from current conditions it may converge to a different equilibrium. In that case, ozone depletion may be irreversible and other atmospheric changes may occur. Similarly, climatic change resulting from POD-induced global warming may be effectively irreversible. Broecker (1987) suggests that oceanic circulation patterns may exhibit diverse equilibria and a global warming could catastrophically shift these circulation patterns, and with them, the climate, to a different stable equilibrium. Returning to the original equilibrium could be difficult or impossible.

A SIMPLIFIED MODEL AND A PARTIAL SOLUTION

Posed in its most general form, the problem is mathematically intractable since it combines stochastic dynamic programming with endogenous learning. However, it is possible to take advantage of specific features of the problem to simplify it in such a way as to provide insight into the current policy question—whether to adopt additional regulations to limit POD use and emissions now, or delay further regulation until we have a better understanding of whether the likely magnitude and consequences of ozone depletion make it imperative.

To simplify the dynamic program, consider the value function currently facing policy makers, where the current time is denoted 0.

$$V_0(s_0, y_0, p_0) = B(p_0, y_{b0}) - \text{Elf}_0 [D(s_0, y_{d0}, x)] - \phi_0 + \delta \, \text{Elf}_0 [V_1^*(s_1, y_1)].$$

Assume that $\text{Elf}_0 [D(s_0, y_0, x)] \approx 0$, since it is not clear whether significant POD-induced ozone depletion has occurred and no resulting damages have been identified. Significant damages are not expected to occur for some time, since many of the potential damages from ozone depletion require cumulative exposure to increased UV flux (for example, human skin cancer). Note, however, that some potential consequences, such as decreased crop yields and adverse effects on aquatic organisms, may result from only one season with increased UV flux if it stresses crops, fish larvae, or phytoplankton during a critical life stage.

Second, assume that the choice of ϕ_0 can be neglected, either because the range from which it can be selected is small relative to the other terms in the value function, or because the research budget is set exogenously and is not subject to policy makers' influence. To the extent that dramatic increases in research may be appropriate, this assumption is not envirely realistic, but ist simplifies the analysis considerably. Moreover, a large share of current research is conducted by industry; government policy makers may have little direct control over the extent and direction of this work.

Using these assumptions, the value function can be simplified to

$$V_0(s_0,p_0) = B(p_0) + \delta \, E|f_0 \, [V_1^*(s_1)]$$

where the notation showing the dependence on y has been suppressed because technological and scientific development is taken as exogenous. For specificity, assume that smaller values of x correspond to states for which ozone depletion is more likely or harmful. In what follows, I drop the time-subscript 0 to simplify notation; the optimal current action is denoted p^*.

This formulation captures the tradeoff between current benefits from using PODs, $B(p)$, and potential future damages, incorporated in $V^*(s_1)$. Since benefits accrue in the current period but damages will not be incurred until sometime in the future, the discount factor δ is a key parameter in the analysis. A higher discount factor (or lower discount rate r, where $\delta = [1 + r]^{-1}$), assigns greater values to future consequences relative to the present; it decreases optimal current production p^*. A lower factor (corresponding to a high discount rate) assigns relatively less weight to future benefits and damages, and so increases p^*.

This formulation also provides insight into the role of uncertainty and learning. If the primary source of uncertainty is whether or not POD emissions will lead to stratospheric ozone depletion, but the consequences of such depletion, should it occur, are comparatively certain, then greater uncertainty in the sense of a smaller probability that ozone depletion will occur increases the optimal p. In this case, the current benefits of POD use are relatively certain but the possible damages are contingent on ozone depletion actually occurring. The less certain policy makers are that such depletion will occur, the smaller are the expected damages, and thus the smaller the current benefits that should be foregone to prevent these damages. If, over time, scientific research increases the perceived likelihood of ozone depletion occurring, the optimal p_t will decline.

In contrast, a reduction in uncertainty about the likelihood and consequences of ozone depletion, holding the expected damages constant, may or may not affect the desirability of adopting additional current emission restrictions. The tradeoff explicit in the value function is between known current benefits and the expected value of future net benefits (benefits less damages). If the expected utility of future net benefits is held constant, the desirability of current regulations is unaffected by the degree of uncertainty. However, if the expected damages, as measured in dollars, are held constant and social utility is a non-linear function of the monetized damages over this range, uncertainty may affect the tradeoff. In the more likely case, society may be risk averse with regard to these damages, so greater uncertainty about the net benefits will be reflected in a larger risk premium and will enhance the attractiveness of additional current emission restrictions. Decreasing uncertainty over time will reduce the risk premium and thus increase the optimal p_t.

The set of possible decisions is effectively constrained, since $p \geq 0$ and the maximum feasible p is limited by demand for the products and processes to which PODs are suited. The lower bound could be relaxed somewhat, if policies to reclaim past POD production from products such as refrigerators and foam insulating boards before it is emitted are considered, but such policies may be very expensive. In any case, only policies from some closed interval [p', p''] need be considered. If the density function $f(x)$, benefit function $B(p)$, and optimal value function $V_1^*(s_1)$ are reasonably well-behaved, the optimal p^* is likely to be a monotonic function of x. That is, if $f^a(x)$ stochastically dominates $f^b(x)$ (simplistically, it assigns greater probability to larger values of x) the corresponding optimal production $p^a > p^b$.

Assume that, for political or other reasons, the choice of decisions is constrained so that only two policies, p^- and p^+, can be considered. For example, p^- could represent adopting a specified set of proposed emission restrictions that would become effective immediately and p^+ could represent maintaining the existing regulations while delaying one period to await the results of a scientific assessment study. Then, p^- will be optimal for all x $< x^*$, and p^+ will be optimal for $x > x^*$. Similarly, if the value of x is unknown but its distribution is $f(x)$, p^- will be optimal for some set of distributions $\{f^-(x)\}$ and p^+ will be optimal for the remaining set $\{f^+(x)\}$. In general, $\{f^-(x)\}$ will include distributions that assign relatively large probability to small values of x, and elements of $\{f^+(x)\}$ will assign relatively large probability to large x values.

If the set of possible distributions $\{f(x)\}$ is suitably limited, it is possible to describe the boundary between the sets $\{f^-(x)\}$ and $\{f^+(x)\}$ by simple constraints on their parameter values. For example, if $\{f(x)\}$ includes only triangular distributions on an interval $[x^-, x^+]$, $\{f^-(x)\}$ will include only distributions for which $E(x)|f(x) < x'$ for some value x', and $\{f^+(x)\}$ will include all distributions with $E(x)|f(x) > x'$. If $\{f(x)\}$ includes a more general set of distributions, the boundary between the sets $\{f^-(x)\}$ and $\{f^+(x)\}$ may involve higher moments of the distribution as well as the expected value.

Using these assumptions, the current policy problem can be represented as the simplified stage of the dynamic program shown in Fig. 5. The current state, incorporating past POD emissions, is s. The possible decisions for the current period are p^-, representing new, additional POD emission restrictions, and p^+, representing no additional regulations in this period. Regardless of the policy chosen, scientific research during the period will allow revision of the current (prior) probability distribution from $f(x)$ to one of two possible (posterior) distributions, $f_1^+(x)$ or $f_1^-(x)$. The first corresponds to learning that potential ozone depletion is not as serious a problem as feared, so the expected value of future benefits and damages conditional on optimal policy thereafter, denoted V^{*1} or V^{*3}, is relatively high. The corresponding optimal decisions in the next period, p^1 or p^3, depend on whether p^+ or p^- was chosen in the current period. If p^- was selected (additional regulations were imposed), the optimal policy p^3 might require relaxing these restrictions; if p^+ was selected (no additional restrictions in the current period), p^1 might correspond to continuing the policy of maintaining pre-existing regulations while monitoring scientific developments.

The second distribution, $f_1^-(x)$, corresponds to learning that ozone depletion is more likely to have serious adverse consequences. The corresponding expected values, V^{*2} and V^{*4}, are comparatively small; $V^{*2} < V^{*1}$ and $V^{*4} < V^{*3}$. The optimal policies in the subsequent period, p^2 and p^4, correspond to imposing further restrictions to limit POD emissions. Because emissions in the current period were not limited along the path leading to V^{*2}, the optimal policy in the next period, p^2, involves more stringent additional emission limitations than the policy p^4 that is optimal if additional emission restrictions were adopted in the current period, so $V^{*2} < V^{*4}$.

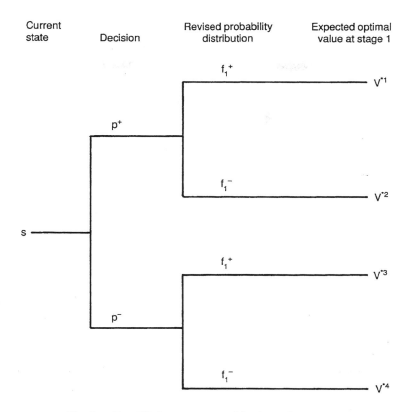

Fig. 5 — Simplified current stage of the dynamic program.

THE CRITICAL PROBABILITY

The optimal decision in the current period is the value of p that maximizes the value function

$$V(s,p) = B(p) + \delta \, E|f \, [V_1^*(s_1)].$$

Because the posterior distribution $f_1(x)$ can be one of only two possible distributions, $V_1^*(s_1)$ is a Bernoulli random variable and $E|f \, [V_1^*(s_1)]$ depends on the subjective probability θ that $f_1(x) = f_1^-(x)$. The expected values of the two candidate policies are

$$V(s,p^+) = B(p^+) + \delta \, [(1 - \theta) \, V^{*1} + \theta \, V^{*2}]$$

$$= (1 - \theta) \, V^1 + \theta \, V^2$$

$$V(s,p^-) = B(p^-) + \delta \, [(1 - \theta) \, V^{*3} + \theta \, V^{*4}]$$

$$= (1 - \theta) \, V^3 + \theta \, V^4$$

where $V^i = B(p^+) + \delta \, V^{*i}$ for $i = 1,2$, and $V^i = B(p^-) + \delta \, V^{*i}$ for $i = 3,4$. V^1, V^2, V^3, and V^4 represent the value of each path at the current stage, assuming optimal decisions at all subsequent stages.

As illustrated in Fig. 6, the optimal choice between the policies depends critically on θ. The figure shows the expected benefits $V(s,p)$ for each policy as a function of θ. For small θ, $V(s,p^+) > V(s,p^-)$ since small θ indicates that ozone depletion is not likely to be a serious problem, so significant emission restrictions are not likely to be required in future periods. For large θ, $V(s,p^+) < V(s,p^-)$ since further restrictions are likely in future periods. The value of θ for which the expected values conditional on each policy are equal is the "critical probability" q.

Figure 6 also illustrates the expected cost of choosing the wrong policy. The expected cost is a linear function of the difference between the subjective probability that emission-limiting regulations will be necessary and the critical probability. When $\theta = q$ the expected cost is zero; for other values of θ it can be calculated as

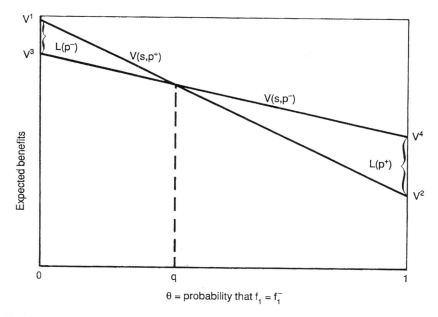

Fig. 6 — Expected benefits of alternative current policies. The lines labelled V(s,p⁻) and V(s,p⁺) show the expected benefits of policies p⁻ (impose additional restrictions on POD emissions) and p⁺ (do not impose additional restrictions at present) as a function of the probability θ that additional emissions restrictions will subsequently be required.

$$E(L) = |B(p^+) - B(p^-) + \delta [(1 - \theta) (V^{*1} - V^{*3}) + \theta (V^{*2} - V^{*4})]|.$$

The costs of choosing the policy that is revealed with hindsight to have been incorrect, $L(p^+)$ or $L(p^-)$, are

$$L(p^+) = - [B(p^+) - B(p^-) + \delta (V^{*2} - V^{*4})]$$

$$= V^4 - V^2$$

$$L(p^-) = - [B(p^-) - B(p^+) + \delta (V^{*3} - V^{*1})]$$

$$= V^1 - V^3.$$

$L(p^+)$ measures the loss if p^+ (no new regulations) is chosen in the current period but substantial emission reductions are subsequently learned to be necessary ($f_1(x) = f_1^-(x)$). Since $B(p^+) > B(p^-)$, the higher benefits from choosing p^+ in the current period partially offset the higher future damages and control costs incorporated in ($V^{*2} - V^{*4}$). Analogously, if p^- (additional emission restrictions) is chosen but these restrictions are subsequently learned not to be necessary ($f_1(x) = f_1^+(x)$), the cost is $L(p^-)$. The foregone benefits in the current period, $B(p^-) - B(p^+)$, may be partially offset by possible increases in future production and emissions reflected in ($V^{*3} - V^{*1}$). $L(p^+)$ largely reflects the long-term control-cost savings possible if emission restrictions begin earlier; $L(p^-)$ primarily reflects foregone benefits in the current period due to unnecessary restrictions.

The critical probability solves the formula

$$(10) \quad \begin{cases} q = \dfrac{1}{1 + \dfrac{L(p^+)}{L(p^-)}} & L(p^-) > 0, L(p^+) > 0 \\[2em] q = 0 & L(p^-) \leq 0 < L(p^+) \\[1em] q = 1 & L(p^+) \leq 0 < L(p^-). \end{cases}$$

It depends on the relative costs of the two possible errors. If either possible cost is not positive the critical probability degenerates to zero or one, and one of the decisions is always better. In the usual case, however, both losses will be positive and q is a function of their ratio. If the ratio $L(p^+)/L(p^-)$ is large, the cost of delaying additional emission restrictions when such restrictions will be necessary in later periods overwhelms the cost of imposing additional restrictions that later prove unnecessary. Thus, q is small, and p^+ is preferred only if policy makers are very sure that subsequent emission restrictions will not be needed. Alternatively, if $L(p^+)/L(p^-)$ is small, the costs of stringently restricting emissions in the future are small relative to the foregone benefits of limiting emissions in the current period, q is large, and additional regulations should not be imposed in the current period (p^-) unless policy makers are reasonably certain that emission reductions will be necessary in the future.

Evaluating the critical probability requires estimating the values of the losses from choosing each policy when, after the fact, the other policy turns out to have been preferred, $L(p^+)$ and $L(p^-)$. These losses in turn depend on the expected values assuming optimal regulations thereafter, V^{*1}, V^{*2}, V^{*3}, and V^{*4}. The solution to the current policy question depends on the expected welfare conditional on optimal regulations in each of the following periods.

Because stochastic dynamic programs require knowing the expected value of future stages to determine optimal decisions, there are two solution methods. If the problem has a finite horizon, one can begin with the final period. The expected values corresponding to each possible terminal state are calculated. From these, one can find the optimal expected values and policies corresponding to each possible state at the penultimate stage. Using the optimal values for each penultimate state, one can solve for the optimal expected values in the preceeding stage, and continue back to the current stage.

If the dynamic program does not have a finite horizon, as in this case, solution is not straightforward. Often, such problems can be solved only if the optimal values and policies are constant or geometrically declining over time. Solution may depend on being able to write the value function and state transition function in simple analytic form.

In order to describe the solution to the current stage of the policy problem illustrated in Fig. 5, I make a series of assumptions that enable me to approximate the optimal values V^{*1}, V^{*2}, V^{*3}, and V^{*4} in the next period. Specifically, I assume that the state s can be adequately represented by cumulative weighted POD emissions, where the weights correspond to the estimated relative ozone depletion efficiencies of each compound. Over the time period considered, this is reasonable because the PODs that are quantitatively important on this basis have estimated atmospheric lifetimes of 75 years or more (Table 1).

Second, I assume that optimal policies beginning in the next period will produce equal cumulative weighted emissions as of a fixed future date, denoted the planning horizon. Regardless of which policy is chosen for the current period, no significant ozone depletion is likely to occur before the horizon so the damages associated with depletion are negligible and do not depend on the current policy choice. The only difference between the expected optimal values in the next stage depend on the costs of reducing future POD emissions: If emissions are not restricted in the current period, achieving the same cumulative emissions through the horizon will require more stringent restrictions in future periods. Since the marginal cost of reductions is presumably increasing and convex, the present value of the cost of achieving a fixed cumulative decrease in emissions through the horizon may increase if the restrictions are constrained to a shorter time period.

A more precise solution would recognize that the optimal level of damages to accept might be greater if POD emissions are not restricted in the current period. Since the cost of achieving a fixed reduction in cumulative emissions through the horizon may increase if current emissions are not restricted, it might be optimal to accept slightly more ozone-depletion-induced damage, and incur slightly smaller control costs in this case. Presumably, the error introduced through this assumption is not large.

Third, to calculate the optimal expected values corresponding to each of the four possible states s_1, I assume the expected value is the same as the value if the optimal cumulative weighted POD emissions through the horizon became known at that time. Thus, f_1^- and f_1^+ might each concentrate their mass at a single point, so that at the next stage it is learned that cumulative emissions should be limited to ψ^- (if the new distribution is f_1^-) or ψ^+ (if it is f_1^+). One way to understand this assumption is to assume that, at the end of the current period, scientific research will yield sufficient understanding of the causes and consequences of ozone depletion that it will be possible to select an optimal level of POD emissions, balancing the costs of depletion against those of emission control. The optimal cumulative emission level will be either ψ^- or ψ^+. A more general interpretation would be that uncertainty about the appropriate level of cumulative emissions through the horizon will be substantially reduced, and that ψ^- and ψ^+ reasonably characterize the possible cumulative emission limits that will appear to be appropriate at that time.

This simplified analysis abstracts from several important features of the problem in order to make it possible to calculate a numerical solution and provide quantitative guidance to policy makers. It assumes that information will arrive early enough that significant adverse effects can be prevented by imposing sufficiently stringent regulations at that time. (Stratospheric ozone concentrations are not sensitive to variations of a few years in the timing of emissions, because of the PODs' long atmospheric residence times.) The only penalty for not imposing additional emission restrictions in the current period is the lost opportunity to distribute any required emission reductions over a longer period, and thereby potentially reduce their cost. Since the environmental consequences do not depend on whether regulations are adopted immediately or not, the analysis reduces to a comparison of the expected economic costs of the alternatives.[3]

Many aspects of the simplified problem are discrete, not continuous as in the real problem. For example, in the simplified problem only two information outcomes are possible: A more realistic subjective distribution for the appropriate level of emission limits would assign positive probability to a continuous range of emission limits. To offset this limitation, the critical probability will be calculated for the entire range of possible cumulative-emission limits, to assess its sensitivity to these choices. Similarly, new information that substantially reduces uncertainty about the ultimate consequences of emissions arrives in a discrete package at a predetermined date. This simplification should bias the results in favor of delayed contingent regulations, because in the real problem information will arrive in smaller bits at irregular, somewhat unpredictable intervals and the uncertainty will never be completely resolved, so one can never avoid the risk of regulating more stringently than necessary.

In the simplified problem, the date and type of new information are independent of the chosen regulatory strategy (learning is passive). In fact, the rate of scientific progress may depend on the regulations chosen: In the extreme case, if emissions are severely limited we might never learn whether significant ozone depletion would have occurred.

The simplified problem also does not account for the potentially irreversible effects that regulation may have on demand and on industry, for the administrative and political costs of developing or revising regulations, or the possibility that advanced notice of impending regulations may reduce the associated transition costs.

Finally, the simplified problem abstracts from the issue of choosing the appropriate level of emissions and corresponding environmental and welfare consequences. In principal, this issue can be solved by comparing the costs of environmental modification with those of emission control, although the consequences and costs of environmental change cannot credibly be measured at present. Despite these simplifications, this simplified problem elucidates many of the issues pertinent to the decision.

The analysis focuses on the critical probability. The results presented in Chapter 10 describe how it depends on the parameters of the decision problem, including the likely severity of the welfare consequences of emissions (reflected in the degree of emission restrictions that are appropriate) and the date at which significant new information will become available. That chapter also describes the sensitivity of the critical probability to alternative assumptions about the likely growth in demand for potential ozone depleters, the elasticity of demand for them, the possibility of future innovation that reduces the cost of limiting emissions, and the discount rate used to compare present and future costs.

PARAMETERS OF THE MODEL

To allow meaningful comparison of the expected costs of regulating now versus waiting for new information, the emission paths corresponding to the two main branches of the decision tree must impose equivalent risk of environmental damage. In the model, alternative regulatory trajectories are assumed to impose equivalent environmental risk if the cumulative weighted emissions of the seven primary potential ozone depleters are equal through the planning horizon.[4] Clearly, cumulative emissions and ozone concentrations under different regulatory trajectories cannot be equal at all dates; however, variations in the timing of emissions on a scale significantly shorter than that of the long atmospheric residence times of the PODs should have little effect on ozone depletion at the horizon and beyond. Differences in ozone concentration before the horizon should also be small (less than 0.1 percent averaged globally) and relatively short-lived (about 10 or 20 years).[5] The associated differences in health and environmental consequences are assumed to be negligible.

Additional emission restrictions in the current period are characterized as beginning at the start of 1988, and additional restrictions imposed in the subsequent period begin at the start of 1995. These dates were selected in mid-1986, based on several considerations. The earliest possible date for new restrictions to be imposed appeared at the time to be January

1988. This date has passed, but in September 1987 the major POD-producing nations signed the "Montreal Protocol on Substances that Deplete the Ozone Layer," which requires specified reductions in POD emissions. The protocol entered into force in 1989, one year after the date of early regulations assumed here (U.S. Congress, Office of Technology Assessment, 1988). In addition, EPA (1987) has proposed rules to implement the protocol in the United States. Delay in initiating emission restrictions of a year should have little effect on the calculated critical probability.

The choice of 1995 as the beginning of the next period allows eight years from the initial decision for the development of new scientific understanding and the design and implementation of regulations, if necessary. If a decision is made to await further information before deciding whether to regulate, one must be prepared to wait several years in order to allow significant scientific progress and to reestablish the political and bureaucratic momentum needed to develop international regulations. Eight years should allow substantial improvement in understanding the environmental consequences of POD emissions, especially considering current programs for measuring ozone and other trace-gas concentrations (National Academy of Sciences, 1984; World Meteorological Organization, 1985). The sensitivity of the results to the choice of the date is tested by repeating the calculations using 2000 as the date at which contingent regulations can be implemented.

The planning horizon is set at 2020 to allow sufficient time for regulations beginning in 1995 to offset unregulated interim emissions growth. Because of the lags between production and emission, and between emission and ultimate effect on ozone concentration, there is substantial delay between changes in production and reversal of any trend in ozone concentration. If the ozone concentration is falling in 1995, even if POD production were sharply reduced or even halted the ozone concentration would likely continue to fall for a few years and would not recover to its 1995 level for perhaps 20 years. The maximum additional depletion over that period should not exceed one percent.[6] Sensitivity of the results to the choice of horizon is explored by additional calculations using 2005 and 2010 as the horizon.

Alternative strategies for immediate regulation are compared with the option of awaiting new information and then regulating only if necessary. For each of these alternatives the critical probability is calculated as a function of the total to which cumulative emissions may need to be limited. The strategies differ in the stringency of regulations in the current period compared with the stringency that would be imposed in subsequent periods if regulations are required then.[7]

CALCULATION OF RESOURCE COSTS AND CRITICAL PROBABILITIES

Calculation of the critical probability for any proposed immediate regulation and cumulative-emission limit requires calculation of the present value of the resource costs of alternative regulatory trajectories (the values V^1, V^2, V^3, and V^4 in Fig. 6). These resource costs are measured as areas under the demand curves for each chemical. The calculations are performed using a computer program that simulates annual demand curves for each commercial application of the PODs over the period 1985 through 2020.

The calculations assume that the regulations will consist of surcharges imposed on POD use. Using a surcharge, the effective price of PODs can be increased so that consumers will switch to alternate products and manufacturers will substitute other chemicals, more conservative processes, or other technological options that become cost effective. The size of the surcharge can be varied to induce the desired amount of emission reductions. The use of such a surcharge induces manufacturers and consumers to adopt the economically efficient set of emission-reducing measures, thereby minimizing the annual resource costs of emission reductions. To minimize the present value of the cost of limiting cumulative weighted emissions, the surcharges should be proportional to the relative ozone-depletion efficiencies of each chemical and rise over time at the discount rate that firms and consumers use in making investment and consumption decisions.[8]

If a surcharge is applied, the resource costs of the regulation can be measured by the area under the demand curve for each chemical between the unregulated price and the price including surcharge.[9] Other regulatory programs, such as marketable permits, mandated control technologies, or selective product bans, could also be applied. These alternative programs would generate the same resource costs, unless they fail to induce the most efficient emission control technologies. In that case, the resource costs would be higher than those calculated.[10] Thus, the assumption that regulation will be characterized by a surcharge is primarily a convenient device to allow calculation of the resource costs.

A total of 28 annual demand curves for specific chemical applications are simulated, for the 36 years from 1985 through 2020. The 28 annual curves measure demand for POD use in 14 applications each in the United States and in the rest of the world. The shapes of the demand curves are based on estimates in Camm et al. (1986) of the chemical prices at which manufacturers would substitute other chemicals or production processes. The curves include the known technological options that are likely to be adopted if the effective price of the PODs were to increase by no more than five dollars per pound (about a 10-fold increase). Camm et al. (1986) focus on technological options in the United States, and discuss the factors that may limit the applicability of their findings to other countries. However, because no better estimates of the demand curves in other countries are currently available, the simulations assume (with one exception) that firms in other countries would respond in the same manner as U.S. firms.[11]

The simulated demand curves shift over time to account for likely growth in demand for chemical use in each application. The likely growth is described in the previous section on production and emission scenarios. In the standard case, demand is assumed to grow along the median projection.

CHAPTER 10
RESULTS AND SENSITIVITY ANALYSIS

The level to which cumulative weighted emissions can be constrained depends on the date at which emission regulations are imposed and the stringency of the regulations. Figure 7 illustrates the effect of these factors. The initial base surcharge, ranging between zero and five dollars per weighted pound, is indicated along the abscissa. In this standard case, POD demand is assumed to grow at the median rates described in the previous section and the surcharge increases three percent per year. The three lines in the figure correspond to regulations beginning in 1988, 1995, and 2000. In the absence of regulations (that is, with a surcharge equal to zero), global cumulative weighted emissions from 1985 through 2020 total about 63.5 million metric tons (Mt). If regulations were to begin in 1988, limiting emissions to 50 Mt would require an initial world-wide surcharge of about $0.90/lb., limiting emissions to 40 Mt would require an initial surcharge of about $1.87/lb., and the minimum attainable level of cumulative emissions, if the initial surcharge were $5/lb., would be about 32.5 Mt.[1] If regulations were not initiated until 1995 larger surcharges would be necessary to limit emissions to the same levels: A 50 Mt limit would require a surcharge beginning at $1.22/lb.; a 40 Mt limit would require a surcharge beginning at $2.83/lb. The smallest attainable cumulative emissions, if regulations did not begin until 1995, would be about 37.2 Mt. If regulations were not imposed until 2000, the range of attainable cumulative emissions is further reduced, and even higher surcharges would be required to hold emissions to any attainable level.

Figure 7 illustrates the apparent inelasticity of the demand for potential ozone depleters: Even with surcharges of several dollars per pound, compared with unregulated prices on the order of 50 cents/lb., simulated cumulative emissions over the next 35 years

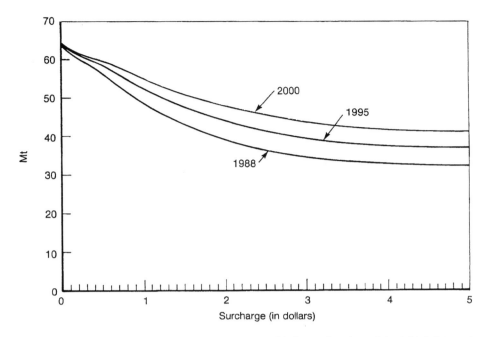

Fig. 7 — Cumulative weighted emissions through 2020 as a function of the initial date and surcharge. POD demand is assumed to grow in accordance with 50th percentile scenario.

fall by no more than half.[2] Limiting cumulative emissions to 35.2 Mt, the level corresponding to continued emissions at 1985 rates (the scenario often assumed for atmospheric-model calculations) would require an initial surcharge of $2.80/lb. if regulations begin in 1988 and would not be possible (under these assumptions) if regulations were delayed until 1995.

In part, the difficulty of reducing emissions reflects the emissions that will occur from the currently existing banks contained in rigid foam and refrigeration equipment. The PODs contained in these banks are not believed to be recoverable at costs less than $5/lb. (Camm et al., 1986). Additional factors include the PODs' superior performance and small contribution to total product price in many applications, requiring large price increases before manufacturers and consumers will substitute alternative chemicals. There is great uncertainty, however, about the technological alternatives that might be adopted if surcharges of several dollars per pound were to be imposed, and consequently about the elasticity of the demand curves. Although current estimates (Camm et al., 1986; Palmer et al., 1980) suggest that substitution possibilities are limited, a surcharge of several dollars per pound would create strong incentives to develop alternative manufacturing processes and products. Consequently, estimates of the minimum attainable emissions using a surcharge of no more than $5/lb. are particularly uncertain.

The resource costs (lost economic surplus) associated with restrictions that reduce POD emissions are substantial. Figure 8 illustrates the present value (in 1985 using a three-percent discount rate) of the resource costs associated with a surcharge beginning at the level indicated on the abscissa and increasing at three percent per year. Using Figs. 7 and 8, one can estimate the resource cost associated with various cumulative-weighted-emission levels. For example, to limit cumulative emissions through 2020 to 50 Mt requires a surcharge beginning at $0.90/lb. if regulations begin in 1988. As shown by Fig. 8, the present value of the associated resource cost is $14.4 billion. If the regulations are not implemented until 1995 the required initial surcharge is $1.22/lb., which is associated with a resource cost of $15.8 billion in present value. These costs are approximately eight times the estimated value of 1985 POD production of $1.8 billion.[3]

Even though a simulated $5/lb. surcharge reduces cumulative emissions through 2020 by about half, it affects the quantity of PODs banked in products much less. As shown by Fig. 9, the simulated weighted bank in 2020 is about 15 Mt, roughly 25 percent less than would be banked if there were no additional regulations.

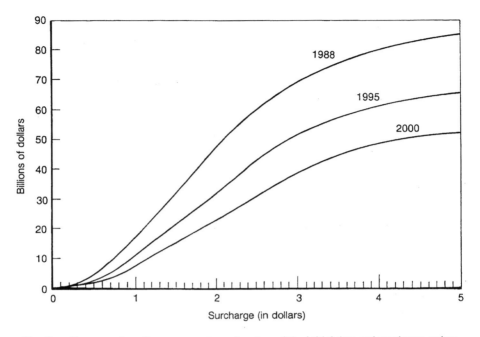

Fig. 8 — Present value of resource cost as a function of the initial date and surcharge, using a 3% real discount rate. POD demand assumed to grow in accordance with 50th percentile scenario. Surcharge increases 3%/yr.

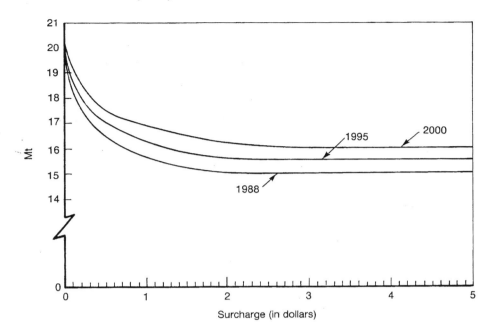

Fig. 9 — Weighted bank in 2020 as a function of the initial date and surcharge. POD demand assumed to grow in accordance with 50th percentile scenario. Surcharge increases 3%/yr.

THE CRITICAL PROBABILITY

Figure 10 illustrates the critical probability that determines whether the expected cost of regulating immediately is greater or smaller than the expected cost of waiting for new information before regulating. The critical probability varies with the cumulative emissions that can be tolerated and with the proposed level of immediate regulations. For example, assume that if research during the current period produces f_1^- as the distribution of effects, optimal regulations will limit cumulative emissions through 2020 to 55 Mt. If the distribution is f_1^+, the 63.5 Mt of emissions that would occur without regulations will be acceptable. The proposed immediate regulations consist of a base surcharge of \$0.30/lb., increasing 3 percent per year. If the new information indicates that cumulative emissions must be limited to 55 Mt, the surcharge will be doubled in 1995 (the path leading to V^{*4} in Fig. 5). If the new information indicates that no regulations are required, the surcharge will be dropped (path 3), and no further costs will be incurred. Alternatively, if the proposed interim regulations are not adopted and the new information indicates that cumulative emissions must be limited to 55 Mt, it would be necessary to impose a surcharge of \$0.79/lb. in 1995 that would increase 3 percent annually (path 2). If the new information indicates cumulative emissions of 63.5 Mt can be tolerated, no regulations would ever be imposed (path 1).

The critical probability can be calculated from the present values of the resource costs associated with each of these four possibilities. Since I do not explicitly evaluate the benefits of POD use, I express these as deviations from the value of the path with no emission restrictions, $V^1 = B(p^+) + V^{*1}$. The values V^1, V^2, V^3, and V^4 are $V^1 - 0$, $V^1 - \$6,785$ million, $V^1 - \$295$ million, and $V^1 - \$6,286$ million. From these, $L(p^+) = \$6,785$ million $- \$6,268$ million and $L(p^-) = \$295$ million $- \$0$. Using Eq. (10), the critical probability

$$q = \cfrac{1}{1 + \cfrac{6785 - 6268}{295 - 0}}$$

$$= 0.36.$$

Thus, if the probability that regulations limiting cumulative emissions to 55 Mt will be

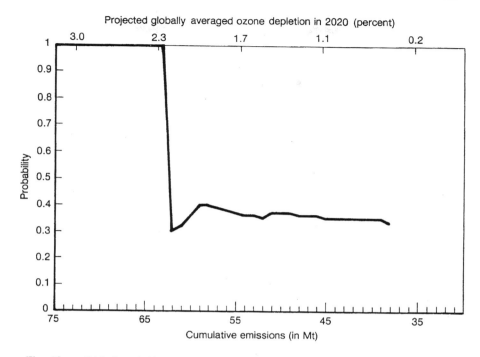

Fig. 10 — Critical probability as a function of acceptable cumulative weighted emissions through 2020: standard case. POD demand assumed to grow in accordance with 50th percentile scenario. Surcharge increases 3%/yr. If surcharge is initiated in 1988 it will be doubled in 1995 if new information suggests POD emissions must be restricted.

required is greater than 0.36, the expected cost of adopting regulations now will be less than the expected cost of awaiting new information and regulating in 1995 only if necessary. If the probability that such regulations will be required is less than 0.36, the expected costs of waiting will be lower than the expected costs of regulating now.

The critical probability varies with the level of cumulative emissions that can be tolerated, but is essentially dichotomous. As shown in Fig. 10, if the level of acceptable cumulative emissions is approximately equal to the unregulated level (63.5 Mt) or higher, immediate regulations cannot be cost-justified (the critical probability is one). If emission reductions may be necessary (to restrict cumulative emissions to 62 Mt or less), the critical probability drops to about 0.35. Over the cumulative-emission range from about 62 to 38 Mt the critical probability is nearly constant.[4] Finally, if the new information might indicate that emissions must be limited to 37 Mt or less, the critical probability cannot be calculated since it is not possible to limit emissions to this level using the assumed demand curves if regulations are delayed to 1995 (Fig. 7). For this standard case, if the level of acceptable cumulative emissions equals or exceeds the unregulated level, immediate regulations are not appropriate. If some emission reductions may be required (about 25 Mt or less), it is cost-effective to wait for better information only if the probability that such reductions will be necessary is less than about 0.35. If larger reductions may be necessary, the critical probability cannot be calculated using the simulated demand curves.

For reference, the abscissa of Fig. 10 also indicates the projected decrease in globally averaged column ozone in 2020 corresponding to various levels of cumulative emissions (indicated along the top of the figure).[5] One way of thinking about the results in Fig. 10 is that by 1995 we will learn either that these depletion estimates are correct or that potential ozone depletion is not an important problem (because it will not occur, or because the consequences will not be serious). Then whether or not immediate regulations are cost-effective depends on the level of potential depletion we can accept. If depletion of 2.3 percent or more is acceptable, immediate regulations cannot be cost-effective, regardless of the probability that ozone depletion will occur. If depletion of less than 0.2 percent is unacceptable, it is not possible to calculate the critical probability without additional information on the costs of reducing emissions more than allowed by the simulated demand curves. Finally, if the acceptable level of depletion is between 0.2 and 2.3 percent, immediate regulations are cost-effective if the probability that ozone depletion will occur exceeds about 0.35.

THE EFFECTS OF ALTERNATIVE PROPOSED CURRENT REGULATIONS

The critical probability depends on the level of interim regulations proposed. In the case illustrated in Fig. 10, the proposed immediate regulations are set so that the surcharge will be doubled in 1995 if the new information indicates emission limitations are necessary. Figure 11 describes the critical probabilities corresponding to a wider set of immediate regulations. The line labelled "0.50" is the same as the line in Fig. 10. The others describe the critical probabilities corresponding to stronger and weaker interim regulations. The line labelled "1.00" corresponds to immediate regulations that are so stringent that, if regulations are needed, it will not be necessary to increase the surcharge in 1995 (except for the three-percent increase that occurs every year). These are the regulations that would be least costly if it were known now that it will be necessary to limit cumulative emissions to some specified level. Each of the other lines corresponds to proposed immediate surcharges that are smaller than the surcharge that will be necessary in 1995, if the new information indicates regulations are needed, by a fixed factor. Specifically, the lines labelled "0.75," "0.50," "0.25," and "0.10," correspond to regulations beginning in 1988 for which the surcharge follows a path from 1988 to 1994 that is smaller than its path from 1995 to 2020 by a factor of 0.75, 0.50, 0.25, and 0.10.[6]

As shown in Fig. 11, the critical probability first falls then rises as the proposed interim regulations become less stringent. The line corresponding to immediate regulations that impose a surcharge only 75 percent as large as the post-1995 surcharge ("75-percent regulations") is below the 100-percent regulation line, and the 50-percent regulation line is even lower. In contrast, the critical probability for immediate regulations that are only 25 percent as stringent as the post-1995 regulations is almost equal to the critical probability for 50-percent regulations, if cumulative emissions may need to be limited to less than 50 Mt. Finally, the critical probability for 10-percent regulations is higher than that for 50-percent regulations; it approximates the 75-percent-regulation line over the interval from about 52 to 37 Mt.

Although one might expect the critical probability to decline with the stringency of the proposed immediate regulations, this is not necessarily the case. Recall from the definition of the critical probability that it depends on the ratio of the costs of choosing each policy if it proves wrong ex-post, $L(p^+)/L(p^-)$. $L(p^+)$ largely reflects long-term cost savings from beginning emission restrictions earlier; $L(p^-)$ primarily reflects foregone near-term benefits because of unnecessary restrictions. As illustrated in Figs. 12 and 13, as the

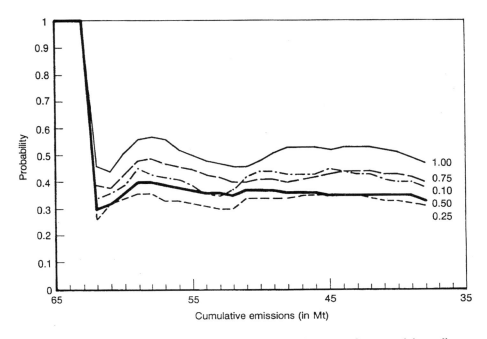

Fig. 11 — Critical probability as a function of the stringency of proposed immediate regulations: standard case. 1.00, Initial surcharge set high enough that no further increase will be required in 1995 if new information suggests POD emissions must be restricted. Other lines, Initial surcharge set at the indicated fraction of this level; e.g., for the line labelled 0.50 the surcharge must be doubled in 1995 if cumulative emission restrictions then appear necessary. See text for details.

proposed immediate regulations become increasingly mild, both terms become small and q may increase or decrease. Figure 11 suggests that the regulations that are most likely to be cost-effective (the critical-probability-minimizing stringency) consist of a surcharge that is about one-quarter to one-half as large as the surcharge that would be appropriate if it were known that it is necessary to restrict cumulative emissions to a specified limit (100-percent regulations).

SENSITIVITY OF THE RESULTS TO THE CHOICE OF PARAMETERS

The following subsections describe the sensitivity of the critical probability to the specific parameter values chosen. As will be shown, variations in most of the parameters have little effect on the results, although variations in the assumed demand growth, dramatic changes in the shape of the demand curves, and large changes in the discount rate can substantially affect the critical probability.

The results illustrated in Fig. 10 represent a standard case. The following subsections illustrate the effect of variations in model parameters by comparing results of alternative calculations to those shown in Fig. 10. Note that the horizontal scale varies across figures.

Alternate Planning Horizon

The calculated critical probability is not sensitive to the choice of 2020 as the horizon through which cumulative emissions and costs are calculated. Figure 14 illustrates the critical probabilities for 50-percent regulations using alternative horizons of 2005 or 2010, as well 2020 (the standard case). Using any of the three horizons, the critical probability falls sharply to about 0.35 if cumulative emissions must be limited to less than their unregulated level (note that the horizontal scales are shifted so that the unregulated cumulative-emission levels are aligned). The lines corresponding to the alternate horizons do not extend as far to the right since the maximum absolute cumulative-emission reduction possible before 2005 or 2010 is much less than the maximum possible by 2020.

Alternate Date of New Information

Figure 15 illustrates the effect on the calculated critical probability of delaying the date at which new information on the severity of ozone depletion will be available. As shown, if the new information will become available only in time to allow regulations beginning in 2000, the critical probability is slightly smaller than if the information will be

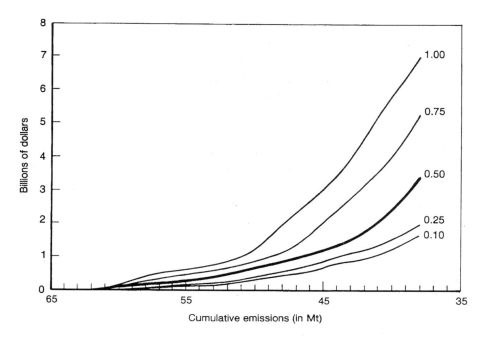

Fig. 12 — Near-term regulatory costs L(p⁻). The present value of resource costs of introducing regulations in 1988 and rescinding them in 1995 as a function of the initial surcharge, corresponding to the fraction of the surcharge needed to limit cumulative emissions to the indicated level.

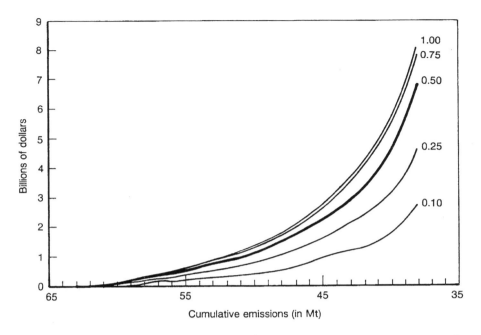

Fig. 13 — Potential long-term cost savings $L(p^+)$. The present value of resource costs of failing to introduce (in 1988) the surcharge corresponding to the indicated fraction of the surcharge needed to limit cumulative emissions if a surcharge must be adopted in 1995, as a function of the acceptable cumulative weighted emissions.

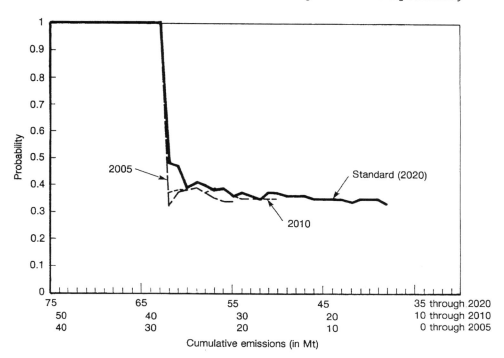

Fig. 14 — Critical probability as a function of the horizon through which cumulative weighted emissions must be limited to the indicated value. Curves are shifted horizontally to align the cumulative emissions that would occur through each horizon in the absence of additional regulations (note alternate scales for abscissa).

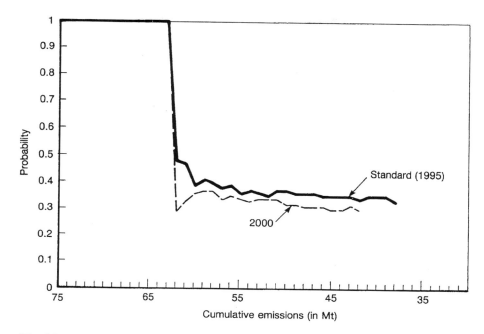

Fig. 15 — Critical probability as a function of the date at which emission-limiting restrictions can be revised.

available in time to implement regulations in 1995. That is, whatever level of emission reductions may be required, if immediate regulations are less costly in the standard case, they will also be less costly if the new information will not arrive until later. Moreover, the delay to 2000 limits the range of emission reductions that can be attained since regulations beginning in 2000 cannot restrict cumulative emissions to less than about 41 Mt. The longer the time required to learn whether ozone depletion problem is serious enough to warrant emission restrictions, the more likely it is that immediate regulations will be less costly.

Alternate Demand Growth Rates

The standard case assumes that POD demand grows in accordance with the median scenario developed in Section A. Figure 16 illustrates the effect of alternate demand-growth scenarios on the cumulative emissions that can be attained and the critical probability. In addition to the standard case, the figure illustrates the critical probabilities corresponding to demand growth at the 25th- and 75th-percentile scenarios. These alternate scenarios represent reasonable but not extreme high- and low-growth outcomes.

The levels of cumulative emissions that can be achieved are significantly affected by the demand-growth rate. If demand grows at only the 25th-percentile rate, cumulative emissions will not exceed 54.5 Mt, even in the absence of regulations (under the standard 50th-percentile growth scenario, limiting emissions to this level would require an initial surcharge of $0.62/lb. if regulations began in 1988). In contrast, if demand grows at the 75th-percentile rate, unregulated cumulative emissions would total almost 73 Mt and even the maximum $5/lb. surcharge would only limit cumulative emissions to 36.6 Mt.

Changes in the expected rate of demand growth shift the critical probability curve horizontally. Expected high growth limits the domain of cumulative emissions over which awaiting new information is less costly, and expands the domain over which immediate regulations are less costly. Expected low growth has the opposite effect. Except for this effect, changes in expected demand growth have little influence on the critical probability. As illustrated by Fig. 16, unless the unregulated cumulative-emission level is acceptable, the critical probability is between about 0.35 and 0.4. If emissions may need to be limited to about 40 to 50 Mt, uncertainty about the future rate of demand growth has almost no effect on the critical probability.

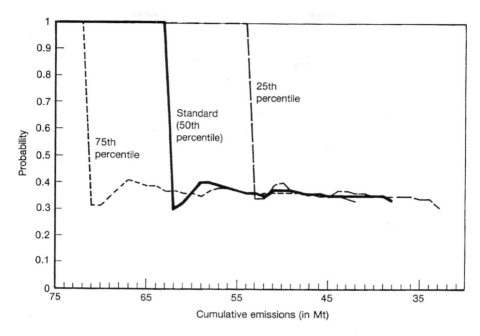

Fig. 16 — Critical probability as a function of future POD demand, for 25th, 50th, and 75th percentile scenarios.

Potential Additional Demand Elasticity

The simulated demand curves are based on analysis of the costs of substitute chemicals, products, and manufacturing processes for various applications. The actual response to regulations is highly uncertain, especially for surcharges of several dollars per pound representing five-fold or greater price increases. Because these prices are so far above current levels, it is extremely difficult to identify the product and process substitutions that might accompany them and to estimate their effects on POD demand.

To assess the sensitivity of the calculated critical probability to the possibility that substantial emission-reducing responses have been overlooked, the simulated demand curves were made uniformly more elastic. Each is multiplied by a constant-elasticity function.[7] The additional elasticity is -0.53 which reduces demand, relative to the standard case, by 50 percent at a $1 surcharge, 64 percent at a $3 surcharge, and 71 percent at a $5 surcharge.[8]

The effect of this substantially increased elasticity is modest. As shown by Fig. 17, the critical probability is only slightly smaller than in the standard case. The additional elasticity also extends the range of emission reductions that can be achieved by regulations beginning in 1988 to a cumulative emission through 2020 of about 15 Mt.

Potential Technological Innovation

Technological innovation may reduce the future costs of limiting POD emissions. Such innovation could include the development of substitute chemicals or alternative products or manufacturing processes that release smaller quantities of PODs. Although these alternatives may not reduce the POD demand at unregulated prices, they could become cost effective at the higher prices associated with regulations. Alternatively, regulations on chemical substitutes for PODs could increase the costs of emission reductions.[9] Such regulations would have a similar effect on the critical probability, but in the opposite direction.

Technological innovation can be simulated by making the demand curves progressively more elastic over time. In the standard case the demand curves expand geometrically over time: The demand at a given surcharge is the same fraction of unregulated demand in every year. With innovation, the development of alternative emission-reducing technologies would further reduce the demand for potential ozone depleters at elevated prices. The specific scenario considered entails significant reductions in the future cost of emission reductions. Under this scenario, innovation affects all

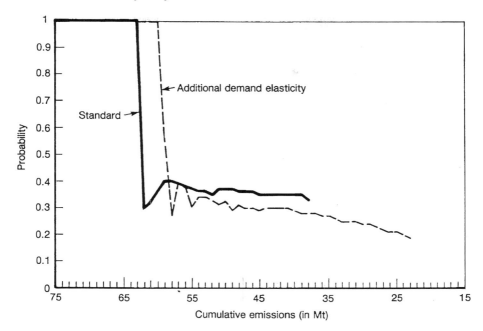

Fig. 17 — Critical probability as a function of POD demand elasticity. The additional demand elasticity line corresponds to a case in which all of the POD demand curves have been multiplied by a constant-elasticity function. Relative to the demand curves in the standard case, POD demand is reduced by about 50% at a $1/lb. surcharge and 70% at a $5/lb. surcharge.

chemicals and applications identically. For every five-year period after 1990 the demand curves for each application are multiplied by a constant-elasticity function. The elasticity of this function increases in each five-year period, to -0.6 by the final period (2016-2020).[10] The effect of this procedure is to make the demand curves in each succeeding five-year interval progressively more elastic. Compared with the standard case, the final-period demand curve is substantially more elastic: At a surcharge of $1/lb. the demand is only 56 percent as large, at $3/lb., 34 percent, and at $5/lb., only 26 percent of demand in the standard case.[11]

The effect of the substantial simulated increase in elasticity over time is similar to the effect of lower demand growth illustrated in Fig. 16. As shown in Fig. 18, expected innovation shifts the critical probability curve to the right. For mild cumulative-emission limits (about 60 Mt) the critical probability rises to 1.0, so immediate regulations cannot be cost-justified. Innovation also increases the emission reductions that can be achieved, thereby increasing the domain on which the critical probability can be calculated down to about 25 Mt. Over the range of intermediate emission limits, the simulated innovation has little effect. Technological innovation also reduces the cumulative emissions that can be achieved, to about 20 Mt in this case (if regulations begin in 1988).

Potential Substitute Chemicals

A dramatic result of technological innovation would be the development of chemicals that substitute for potential ozone depleters in a wide range of applications. Several chemicals that might substitute for the primary PODs have been synthesized, but commercially feasible production processes have not been developed. For example, it is believed that CFC-134a could substitute for CFC-12 in nearly all refrigeration applications and CFC-123 or CFC-141b could potentially replace CFC-11 in rigid-foam applications.

Figure 19 illustrates the effect that the development of general chemical substitutes ("backstop" chemicals) could have on the critical probability. Specifically, it assumes that substitutes for all seven PODs are available at weighted surcharges of $2/lb. or $5/lb. The simulated demand for all seven PODs falls to zero when the backstops become cost-effective. As shown, the possible development of substitute chemicals has a significant effect on the critical probability and extends the range of achievable cumulative-emission reductions. The existence of backstop chemicals increases the critical probability over a range of moderate emission reductions, the location of which depends on the price at which

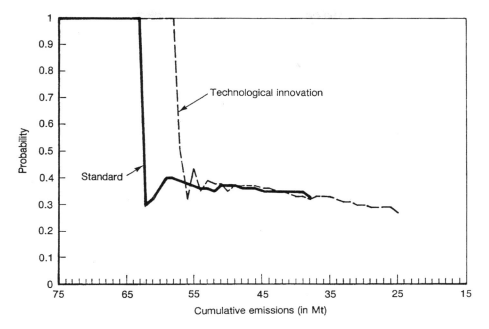

Fig. 18 — Critical probability as a function of technological innovation. The technological innovation line corresponds to a case in which all of the POD demand curves become increasingly elastic over time. Relative to the demand curves in the standard case, POD demand in the period 2016-2020 is reduced by about 44% at a $1/lb. surcharge and 74% at a $5/lb. surcharge.

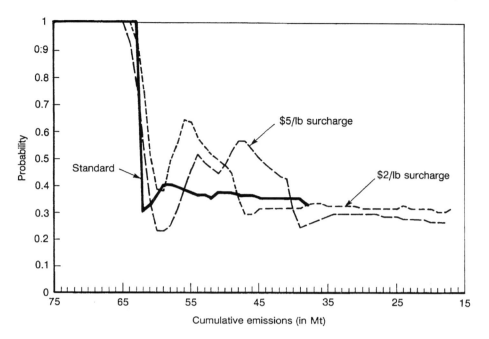

Fig. 19 — Critical probability as a function of the price at which substitute chemicals become available. The alternate lines correspond to the existence of perfect chemical substitutes for all PODs that become cost-effective at surcharges of $2/lb. or $5/lb., weighted by relative ozone depletion efficiency. Simulated POD demand is zero at surcharges above these levels.

the substitutes become available. For more stringent emission' reductions, the critical probability is slightly smaller than in the standard case. If only modest cumulative-emission reductions may be necessary (to about 55 Mt), the existence of general chemical substitutes would make the strategy of waiting before adding regulations more attractive. For larger reductions it has little effect.

Linear Demand Curves

The simulated demand curves in the standard case produce a relationship between the surcharge and cumulative emission reductions with nearly constant elasticity (about equal to -0.3). As a result, the elasticity of the resource costs associated with each path of Fig. 5 with respect to the cumulative-emission limit are nearly the same and the critical probability is not sensitive to the exact emission limit. As noted above, it is difficult to estimate the shape of the demand curves at chemical prices far about current prices. Large surcharges would provide a strong incentive to develop substitute chemicals, manufacturing processes, and products. Thus, even if backstop chemicals are not available the demand curves could become more elastic at higher prices.

A simple alternative to the standard simulated demand curves is to assume the demand curves for all seven PODs are linear. For comparison with the standard curves, simulated demand under the linear functions is about half as large at a $5/lb. surcharge as at no surcharge, about the same as for the standard demand curves. As shown by Fig. 20, the critical probability calculated from the linear demand curves is qualitatively different than in the standard case: It declines monotonically as greater emission reductions may be necessary. At the smallest achievable cumulative emissions it is about equal to the standard-case critical probability but for less stringent reductions it is substantially larger. Thus, if the demand curves exhibit substantially greater elasticity at elevated prices the critical probability may be higher than in the standard case (favoring delayed regulations) and may depend on the level of emission reductions potentially required.

Alternate Discount Rates

The choice of an appropriate discount rate to use in public decision making is a confusing and contentious issue (Goodin, 1982; Lind et al., 1982; Quirk and Terasawa, 1987; Kolb and Scheraga, 1990). The three-percent real rate used in the standard case is supported by the argument that the discount rate should approximate the long-term general

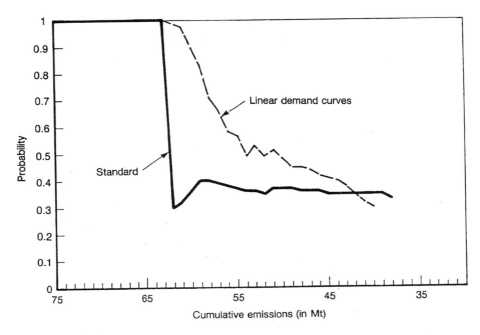

Fig. 20 — Critical probability corresponding to linear POD demand curves. In the alternate case, demand curves for all PODs are linear with demand at a $5/lb. surcharge equal to half the value at zero surcharge.

economic or productivity growth rate, or the social rate of time preference. Unfortunately, as discussed in Chapter 9, the choice of discount rate can be expected to influence the critical probability, unlike many of the other parameters considered, as illustrated by Fig. 21.[12]

The standard case uses a real discount rate of 3 percent. (All prices in the simulation are real.) The critical probability calculated without discounting (that is, using a 0 percent rate) is uniformly lower than in the standard case. Similarly, the critical probabilities calculated using higher rates (6 and 10 percent) are uniformly higher. Using a real discount rate of between 0 and 6 percent has little effect on the results: The critical probability is relatively constant over the relevant domain and its value is between about 0.3 and 0.45. Using a 10 percent rate, whether it is cost-effective to impose regulations in the current period depends on the extent of emission reductions that may be appropriate. Except for stringent emission limits (45 Mt or less), the critical probability is much higher than for lower discount rates.

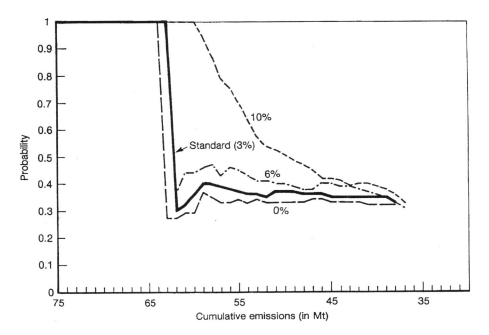

Fig. 21 — Critical probability as a function of real discount rate. In each case, the surcharge increases at the same rate as is used to calculate the present value of resource costs, so the surcharge trajectories always minimize the present value of the costs of limiting cumulative weighted emissions to the indicated level.

CHAPTER 11
CONCLUSIONS

The policy question of what response to take to current and developing understanding of the possibility that emissions of certain anthropogenic chemicals may deplete stratospheric ozone and promote a global warming, thereby increasing ultraviolet radiation at the earth's surface, modifying climate, and potentially producing a host of adverse effects, can usefully be structured as a stochastic dynamic program with endogenous learning. This structure clarifies the relationships between appropriate actions and understanding of the relationships between policy and possible outcomes. However, a structure that incorporates all of the prominent features of the problem is mathematically intractable. To provide useful policy guidance, a substantially simplified model like the one presented may be valuable.

The results calculated using such a simplified model are striking. They suggest that whether immediate regulations to reduce the risk of stratospheric ozone depletion and global warming are justified by an expected-resource-cost analysis depends almost entirely on the degree to which emissions may need to be restricted, and on the discount rate used to compare costs incurred at different times. The results are insensitive to substantial variations in most of the other parameters.

The model allows calculation of a "critical probability" that characterizes the conditions under which the insurance benefits of immediate regulations exceed their cost. This critical probability can function like the standard of proof required in judicial settings. Like the standard of proof, it specifies the degree of confidence policy makers must have that POD emissions will need to be restricted to avoid significant adverse ecological effects in order for them to judge that additional emission restrictions should be adopted at present. If policy makers' perceived probability that emission reductions will be required is greater than the critical probability, the strategy of adopting regulations immediately will impose lower expected resource costs; if the probability is lower, waiting for improved understanding of the likelihood and consequences of ozone depletion before acting will be cost-effective. This section has not addressed how to estimate this probability, but it is clear that such an estimate should consider the best available scientific evidence on the likely extent of future ozone depletion, global warming, and their consequences.

Characterizing the results as a required standard of proof that policy holders must believe is met in order to support additional regulations may significantly reduce the burden on policy makers. The conventional decision-analysis approach would require policy makers to assess their complete subjective probability distributions for the extent and consequences of future ozone depletion. This distribution would be used to integrate the value of alternative outcomes across branches of a decision tree, and the output would consist of expected values corresponding to alternative policies. With this approach, the role of the subjective probability judgments and the sensitivity of the policy choice to variations in these judgments would be concealed through the integration. Sensitivity analysis would require recalculating the expected values for each probability distribution.

In contrast, the critical-probability approach focuses attention on the subjective probability judgment and makes its role in the conclusions transparent. This approach does not require scientists and policy makers to develop a complete distribution, but only to assess whether the bulk of the distribution lies to one side or the other of a specified cut off. It reduces the level of agreement needed to obtain a consensus for policy and clarifies the beliefs that require agreement. Policy makers may be reluctant to be publicly identified with a precise distribution, but much more willing to state whether they believe the evidence is sufficiently convincing or not. Because assessing a complete subjective probability distribution is time consuming, unfamiliar, and not well done by many policy makers (Kahneman et al., 1982), this alternative to the conventional approach may be of value.

A possible disadvantage of this approach is that it convolutes outcome and value uncertainties in judgments about the critical probability. The approach requires policy makers to determine whether the likelihood of effects that are sufficiently adverse to merit imposing the costs of measures to reduce emissions exceeds the critical probability. It may be possible to disaggregate the outcome and value uncertainties by assessing values for the possible consequences and using these to convert the critical probability to a value that corresponds only to scientific uncertainty about the processes leading to ozone depletion and consequences in the biosphere.

The application presented here is particularly simple, because I have considered only Bernoulli probability distributions, for which a single probability characterizes the entire distribution. But the approach could be extended to more general distributions. If it were, policy makers might have to make a more difficult judgment than one about the level of a single probability, but they would only have to determine whether their own distributions corresponded to one or another subset of the admissible distributions. In many cases, this should be an easier task than specifying the shape of the entire distribution.

Sensitivity analysis for this problem shows that, over a wide range of assumptions, the critical probability is a nearly dichotomous function of the extent of emission reductions that may be necessary. If the cumulative emissions that will occur in the absence of additional regulations will not produce significant adverse environmental changes, immediate regulations cannot be cost-effective. If emission reductions may be necessary, the critical probability falls between about 0.3 and 0.5 over the domain of cumulative-emission limits for which it can be calculated.

The assumptions about POD demand do not allow calculation of the resource costs of eliminating emissions. The demand curves only include responses that would occur at price increases of less than \$5/lb. (a nearly ten-fold increase in the current prices of CFC-11 and 12). Under the standard-case assumptions, price increases of no more than \$5/lb. reduce cumulative weighted emissions through the 2020 planning horizon by half, to about the level that results from continued emissions at current rates. Development of substitute chemicals or other products or processes that reduce the cost of limiting emissions extends the range of emission reductions for which the critical probability can be calculated.

The calculated critical probability is not sensitive to reasonable variations in most of the parameters. Advancing the planning horizon (the date from which cumulative emissions under the delayed regulation strategy equal those achieved by immediate regulations) from 2020 to 2005 has almost no effect on the critical probability. Delaying the date by which regulations based on new information can become effective from 1995 to 2000 only slightly decreases the critical probability, making immediate regulations more likely to be favored; advancing the date of new information would make immediate regulations less attractive. Changes in assumed growth of demand for these chemicals affect the critical probability, in that immediate regulations cannot be cost-effective if unregulated emissions will not exceed the acceptable level, but otherwise have no effect. Substantially increasing the demand curves' elasticity, in all periods (to reflect additional consumer response or other emission-reducing measures) or progressively (to reflect technological innovation), has almost no effect on the critical probability, and assuming the existence of general substitutes for all seven PODs at price increases of $2 to $5 per weighted pound only shifts the critical probability by 0.1 or 0.2 for certain cumulative emission limits. In contrast, assuming that the demand curves exhibit markedly increasing elasticity at higher prices (as do linear demand curves) fundamentally changes the critical probability. Instead of remaining constant over the domain of emission reductions for which it can be calculated, it falls almost monotonically with increasing emission reductions. Finally, the choice of discount rate used to compare current and future costs can affect the critical probability. Discount rates between zero and six percent have relatively little effect (shifting the critical probability by about 0.1 or less) but a rate as high as 10 percent has an effect similar to assuming linear demand functions: It substantially increases the critical probability for relatively moderate cumulative-emission reductions (less than about 15 Mt) and makes the critical probability sensitive to the choice of cumulative-emission limit.

If immediate regulations are cost-effective, the choice of appropriate stringency remains. The optimal stringency depends on the probability that restrictions will be necessary. For a specified optimal emission limit (conditional on learning that emissions must be reduced) the level of immediate reductions that is most likely to be cost-effective is modest, corresponding to surcharges perhaps one-quarter to one-half as large as the surcharge that would be appropriate if it were known that emission reductions would be required (Fig. 11). If it might be necessary to reduce projected cumulative emissions through 2020 by 20 percent (to about 50 Mt), a reasonable set of immediate regulations would be equivalent to a surcharge beginning at about $0.25 to $0.50 per pound. If more stringent reductions may be necessary, to 40 Mt, a reasonable surcharge would begin at $.50 to $1.00 per pound.

The simplifying assumptions adopted to make the analysis tractable limit its realism in several important aspects. Perhaps most important is that it starts with the assumption that the level to which cumulative emissions may have to be limited is known. In reality, the choice of an appropriate emission level depends on the relative costs of the environmental damages resulting from emissions and the costs of reducing emissions. It may also depend on how the costs and benefits of regulations are distributed between and within nations. However, it is not possible, at present, to reliably calculate the costs of environmental changes due to potential-ozone-depleter emissions, due to both outcome and value uncertainties. The quantitative relationship between emissions and ozone change depends on complex physical and photochemical atmospheric processes that are not completely understood, on natural and anthropogenic emissions of other gases, and on other factors. The relationships between stratospheric ozone concentrations and effects on humans, plants, animals, and materials are not well understood, and important effects may not yet be identified. Quantitative descriptions necessary to estimate incremental costs are rare. Finally, assigning values to effects on human health and on the global ecosystem is notoriously difficult.

Fortunately, given the difficulty in choosing an appropriate level of cumulative emissions, the critical probability is not sensitive to the exact level under most of the assumptions considered. Neither is it sensitive to the exact level of regulations proposed: Wide variations in the proposed initial surcharge yield similar critical probabilities (Fig. 11).

The numerical analysis assumes that all uncertainty about the acceptable level of cumulative emissions will be resolved within a few years. In fact, uncertainty will endure. This simplification enhances the attractiveness of delaying regulations in the model and thus increases the calculated critical probability. In the real world, unlike the model, even if we delay the decision of whether to regulate, the risk of making the wrong choice remains.

The analysis is also limited by the available data on the shape of demand curves, especially outside the United States. In large measure, the simulated demands outside the United States mimic those within. Improved information on the alternatives to potential ozone depleters that would become cost-effective at higher prices is also needed, since variations in the demand curves reflecting backstop chemicals or increasing elasticity at higher prices can affect the critical probability. The simulated demand curves do not allow estimates of the costs of substantial reductions (e.g., more than about 50 percent) in POD emissions.

The analysis has not addressed other important issues associated with the decision of whether to adopt additional emission-limiting regulations in the near future. These include possible effects on industry of imposing emission-limiting regulations that cannot be reversed if regulations are removed and potential reductions in the transition cost of regulations if they are preceded by substantial advance notice. Both effects would tend to make immediate regulation less attractive. In contrast, if a strategy of waiting for improved information before deciding whether to regulate is adopted, it may be wise to invest more in research and monitoring to hasten development of that information. Since at least some of the anticipated consequences of reduced ozone concentrations (e.g., human skin cancer, cataracts, materials damage), may take many years to manifest, additional research to speed recognition of adverse effects, if they occur, can be valuable. If the additional costs of this research were added to the expected costs of the await-better-information strategy, it would reduce the critical probability and make the immediate-regulation strategy more favorable.

Potential ozone depletion and climatic change are global issues: Their effects, if realized, will be felt world-wide. This section explicitly avoids the important issues associated with the coordination of action among nations. It focuses instead on the logically prior question of whether, from a global perspective, immediate regulations may be appropriate. The results suggest that whether immediate regulations are cost-justified depends primarily on the quantity of future emissions that is acceptable and the likelihood that regulations to limit emissions to that level will be necessary.

PART II

VALUE UNCERTAINTY:
FOOD-BORNE RISK

CHAPTER 12
VALUING HEALTH RISKS

Changes in the physical environment are frequently difficult to value. Such changes may affect human welfare through modification of factors that affect human health or the utility derived from various activities. Improved environmental quality may create opportunities for additional outdoor recreational activities, or may increase enjoyment of current activities by providing better visibility, reduced noise, cleaner water, or other amenities that affect enjoyment of these activities. In addition, people may simply value the existence of certain environmental conditions for philosophical, religious, or other reasons.

The forms in which the benefits of environmental policies may be realized include improved opportunities for recreational and other activities, aesthetic effects such as improved visibility and reduced noise, direct effects on welfare such as reductions in human-health risks, and others. Because these are often not individually traded through markets, conventional cost-benefit techniques of valuing changes at their market prices are not applicable. Two classes of methodologies for valuing these benefits have been developed. The revealed-preference methodology estimates values by observing market behavior related to the condition and inferring values from it. For example, information about the value of improved recreational opportunities can be inferred from people's expenditures on travel or equipment needed to participate in these activities, and information on the value of changes in health risks can be inferred from differences in wages between occupations with differing risks. The contingent-valuation method is more direct. It simply asks people to report their willingness to pay (WTP) for specified improvements in environmental conditions, or willingness to accept (WTA) compensation for specified decrements. The valuation is called "contingent" to emphasize that these values depend on the specific changes posed.

In this section, both methods are applied to estimate the benefits of reducing a certain set of health risks, those that arise from residual pesticides on fresh produce. Changes in health risks are among the most troubling policy consequences to value. Although changes in the probability of diverse health outcomes are implicit in many market transactions, health risks are not explicitly traded. Moreover, health protection has symbolic benefits and risks to health are among the policy outcomes that people are most resistant to measuring in economic terms; non-economists are apt to view health as priceless and efforts to explicitly value it as immoral (Asch, 1988; Fuchs and Zeckhauser, 1987; Kelman, 1981).

Despite these difficulties, policy decisions must be made, and resource allocations inevitably reveal the values implicitly assigned to various health risks. The notion that such valuations can somehow be avoided is at best a comforting fiction; maintaining it can be expected to produce inefficient resource allocations, since greater health benefits could be achieved at the same price if valuations were made explicit.

Government is frequently involved in safety regulation, and this involvement is generally recognized as appropriate. Thus, U.S. government agencies regulate automobile design, airplane design and flight practices, nuclear power plants, workplace conditions, a wide variety of consumer products, air and water quality, and food products. One justification for such governmental involvement is that information about these risks is a public good that the government can efficiently collect and synthesize. Because the number of risks is great, individuals cannot be expected to obtain and evaluate enough information to make informed decisions about many of the risks they face.

Determining whether a risk is acceptable requires more than assimilation and evaluation of information about possible health outcomes and their probabilities; it also requires comparison with the benefits of the risky activity. Although some analysts would argue that the government should exercise paternalistic judgment in making this comparison, many authors recommend that the government set safety standards to reflect individual preferences. That is, the standards should induce consumers to make the decisions they would make freely, if their decisions were informed.[1] If the government is to do so, it must estimate the costs and benefits that consumers assign to various products and activities.

I describe a pilot study to estimate the value that consumers assign to reductions in health risks and to learn more about how people estimate, evaluate, and manage risks to their health. The study examines consumer choice in a familiar context: selection of fresh produce, specifically, consumers' choices between conventionally and organically grown produce. Although most American consumers purchase conventionally grown produce, some are sufficiently concerned about the possible health risks from residual pesticide contamination and other aspects of modern agricultural practice that they choose to pay a higher price to obtain organically grown produce—fruits and vegetables that are grown without direct application of pesticides or other synthetic chemicals. Even among the majority who do not purchase organically grown foods, about 75 percent report that they consider the use of pesticides, herbicides, additives, and preservatives a serious hazard (Food Marketing Institute, 1986). What do consumers' choices between organically and conventionally grown produce tell us about their valuation of reductions in personal health risk?

METHODS FOR VALUING HEALTH RISKS

The choice between organically and conventionally grown produce provides an opportunity to compare estimates using each of the primary valuation methodologies, revealed preference and contingent valuation. Each method has strengths and weaknesses. Reliable estimates of the value of risk avoidance may require contributions and cross-validation from both methods.[2]

The revealed-preference approach uses consumer decisions in an actual market where the commodities traded differ in price, inherent risk, and possibly in other dimensions. The principal assumption of this approach is that consumers choose particular commodities by optimally trading one valued attribute for another.[3] For example, given two otherwise similar jobs that differ in risk, workers will not accept the more dangerous job unless they are offered a higher wage in compensation. Hedonic regression techniques are often used to estimate the implicit price for each attribute of the commodity, that is, the additional cost needed to procure a unit with a specified improvement in only one dimension. Most revealed-preference studies of risk taking have analyzed cross-sectional differences in wage rates associated with varying job classes or differences in property values associated with different levels of air pollution or other attributes, but the technique has been applied to a broad range of other problems.[4] The travel-cost method, which measures the value of a resource by the effort people expend to use it, is a variant of this approach. Other studies of food-borne risk have employed time-series analyses to estimate preferences from changes in market behavior when new information that affected the perceived risk of specific foods was released (Foster and Just, 1989; Johnson, 1984; Shulstad and Stoevener, 1978; Swartz and

Strand, 1981); Ippolito and Ippolito (1984) use a similar approach to analyze trends in cigarette smoking.

The contingent-valuation approach is more direct: One simply asks a sample of representative individuals how much they value a carefully specified change in risk or some other condition. A criticism of this method is that it is difficult for an individual to quantitatively estimate a value for something he cannot purchase in a market, at least not without also purchasing an associated bundle of other qualities.[5] Moreover, economists are often skeptical of contingent-valuation survey results, suspecting that respondents reply strategically, that is, misrepresent their preferences in order to influence the outcome of the exercise. In related contexts, a substantial literature has developed concerning incentive-compatible revelation mechanisms, to which one's best strategic response is his honest response.[6]

The results of most contingent-valuation studies suggest that the possibility of strategic responses is not a significant problem, perhaps in part because it may be difficult for an individual to calculate his optimal strategic response. The tenuous connection between survey responses and ultimate policy further limits the utility of a strategic response. While it is admittedly difficult for an individual to accurately estimate the value of a risk reduction, change in health status, visibility, or other non-traded attribute, many contingent-valuation studies have produced what appear to be reasonable results, and comparisons between contingent-valuation and revealed-preference estimates show reasonable agreement.[7]

I estimate individuals' willingness to pay for risk reductions using each methodology. Using the revealed-preference method, comparable prices for organically and conventionally grown produce are used to estimate incremental willingness to pay for organically grown produce. Using the contingent-valuation methodology, consumers are asked to state their incremental willingness to pay. I estimate the risk avoided by consuming organic instead of conventional produce and combine this with the estimated willingness to pay to calculate an implied "value of life." This implicit value of life is a convenient, widely used summary statistic to describe willingness to pay to reduce mortality risk. It is defined by the formula

$$\text{Implicit value of life} = \frac{\text{Willingness to pay to reduce risk}}{\text{Incremental mortality risk}}$$

which implies that the amount one should be willing to pay to avoid an incremental risk is equal to the value one implicitly places on his or her life divided by the risk increment. To minimize the cost of achieving a given reduction in overall risk, one should allocate resources across risks so as to equalize the value of life implied by each.[8]

The implicit value of life is illustrated in Fig. 22. It is reasonable to expect WTP to be an increasing, convex function of the incremental mortality risk as illustrated (Jones-Lee, 1974). For a very large risk, WTP may be limited only by the individual's resources, assuming the marginal utility of those resources is much reduced if he or she does not survive the risk. The compensation demanded to accept a large risk (willingness to accept compensation or WTA) may be infinite. For small risks the difference between WTP and WTA should in theory be negligible, since they differ only by the income effect corresponding to whether one is endowed with the risk or not. The implicit value of life is defined only for these small risks, and represents the slope of the WTP curve in this region.[9]

Although economic theory suggests that WTP and WTA should be virtually equal for the small risks typically considered in environmental policy making (annual mortality risks of about 10^{-4} or less), empirical studies have typically found people to report a much higher WTA than WTP for comparable risk increments or other attributes (Bishop and Heberlein, 1979; Knetsch 1989, 1990; Knetsch and Sinden, 1984; Mitchell and Carson, 1986; Gregory, 1986; Gregory and McDaniels, 1987). The difference is often attributed to psychological factors such as those incorporated in prospect theory (Kahneman and Tversky, 1979). Coursey et al. (1987) suggest that the disparity can decrease if consumers learn more about the outcome they are valuing and gain experience in making the valuation through a repeated auction mechanism, although Knetsch (1990) claims that substantial differences still remain. Hanemann (1984b) suggests the availability of substitutes can have a major effect on the difference.

Estimates of incremental WTP for organically grown produce can be readily obtained using both revealed-preference and contingent-valuation methods, since these are traded in markets. Estimating WTP for reductions in health risk is more difficult, however, because of substantial outcome uncertainty: The magnitude and character of the health risks avoided by substituting organically grown produce for its conventional counterpart are not known. As described in Chapter 15, knowledge of the toxicological effects of residual pesticides on humans is inadequate to reliably estimate the health risks. The organically grown product may even pose offsetting risks that outweigh those of the pesticide residues.

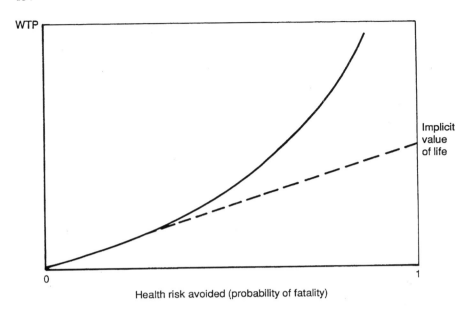

Fig. 22 — Theoretical willingness to pay for risk reductions and the implicit value of life.

To handle this outcome uncertainty, I use two types of risk measures. First, I develop a set of "risk indices" that attempt to characterize the relative risks from pesticide residues across foods and estimate the amount that consumers are willing to pay to reduce the risk as measured by these indices. The risk indices are based on average concentrations of pesticides in foods, where the concentrations are weighted by alternative indices of their potential toxicity. Second, I develop several rough estimates of the cancer risk due to consuming conventionally grown produce using dose-response information for several supposed carcinogens. The approximate value of life is derived using these estimates.

OUTLINE

The following chapter provides background on pesticide use, the risks associated with pesticide residues, and organically grown produce. The chapter also develops a simple theoretical model of the consumer's choice between organically and conventionally grown produce that serves as the basis for the empirical analysis that follows. Finally, the chapter describes consumer understanding of the risks associated with organically and conventionally grown produce and the considerations involved in the choice between them, based on a series of focus-group interviews. Chapter 14 describes the price data and statistical models used for estimating revealed incremental WTP for organically grown produce and contrasts reported WTP assessed using contingent valuation. Chapter 15 describes the construction of risk indices to measure the difference in risk between organic and conventional produce, the estimated lifetime cancer risk from pesticide residues, and consumer-reported estimates of the magnitude of these risks. In Chapter 16, the price and risk information are combined to produce revealed-preference and contingent-valuation estimates of WTP for risk reductions, and information on consumer management of other health risks is evaluated for consistency with these valuations. Chapter 17 suggests policy implications and directions for future research.

CHAPTER 13
CHOOSING AMONG POTENTIALLY HAZARDOUS FOODS

This chapter presents background information on conventionally and organically grown produce. Subsequently, a simple model of consumer choice between these products is developed and information on consumers' understanding of the risks and considerations involved in the choice, as reflected in focus-group discussions, is presented.

PESTICIDES AND OTHER FOOD-BORNE HEALTH RISKS

Modern agriculture depends on the extensive use of pesticides to obtain high yields and low growing costs. Nearly 50,000 formulations are registered for use with the U.S. Environmental Protection Agency (EPA) (Moses, 1983). A relatively small number of pesticides account for the majority of use, however. According to Smith (1977), 31 insecticides account for about 93 percent of the insecticide market, 37 herbicides account for 98 percent of the herbicide market, and 19 fungicides account for 94 percent of the fungicide market. In the United States alone, more than one billion pounds of insecticides, herbicides, fungicides, acaricides, nematicides, and fumigants are applied to conventional crops annually, although about half this total is applied to nonfood crops such as cotton and tobacco (Pimentel et al., 1978; McEwen, 1978).

Pesticide use varies widely by crop and region. On average, 17 percent of the acreage devoted to agriculture (excluding pastures) is treated with herbicides, 6 percent with insecticides, and 1 percent with fungicides. Application to fruits and vegetables is much higher, however. Sixty-nine percent of the acreage used in growing conventional fruit is treated with insecticides, including 91 percent of apple and 72 percent of citrus acreage (Pimentel et al., 1978). Pesticides are also used on lawns and gardens and to fumigate buildings.

Such widespread use may present serious risks. Public attention is periodically drawn to a newly recognized pesticide hazard, such as the use of EDB as a grain fumigant (now banned; Johnson, 1984) or the contamination of milk by heptachlor (applied to pineapple plants, parts of which are used as dairy-cattle feed in Hawaii; Foster and Just, 1989). Other times, an isolated instance of pesticide misuse can cause widespread illness or death and disrupt markets as people avoid the potentially tainted product. Such an instance occurred in July 1985, when watermelons grown in a few California fields were found to be contaminated by aldicarb. Scores of people became ill and the state required the destruction of all watermelons then in retail stores. In 1989, widespread publicity about the possible carcinogenicity of a growth regulator used on apples (daminozide or Alar) resulted in substantial reductions in apple and juice sales, withdrawal of apples from school lunches, and other disruptions.

Pesticide manufacturers are required to register their formulations with EPA. The agency sets legal "tolerances" for each food product on which the pesticide may be used that define the maximum concentration of the pesticide that may remain in the food when it is sold to the consumer. The U.S. Food and Drug Administration and many state agencies have active testing programs to ensure that pesticide levels do not exceed these tolerances, but because of the time and expense involved only a small fraction of food crops are tested. The relationship between these tolerances and the quantities consumers ingest is not clear, because peeling, washing, and cooking produce removes or degrades some of the pesticides that remain.

Humans are exposed to pesticides by other routes, in addition to ingestion of residual quantities in food. Pesticides are an important source of groundwater pollution (Assembly Office of Research, 1985). Pesticide intake from drinking water and through the atmosphere may rival dietary intake, and pesticides in water can be concentrated by fish and plants that are ultimately consumed by humans (Coffin and McKinley, 1980). Pesticide production, transport, and application is hazardous to workers and bystanders. The outstanding pesticide-related tragedy was the December 1984 release of methyl isocyanate from Union Carbide's Bhopal, India plant, causing the deaths of over two thousand local residents. The methyl isocyanate was produced for use in manufacturing aldicarb. Subsequently, a release from Union Carbide's West Virginia plant injured 135. The World Health Organization estimates that 500,000 people are poisoned annually by pesticides; for about 5,000 the exposure proves fatal (Moses, 1983). Among those most at risk are farm workers, many of whom are migrants, and their children who accompany them to the fields.

Although the risks of acute pesticide poisoning are relatively well understood, the risks posed by long-term exposure to low levels are more speculative. Tests for long-term effects are incomplete, but several pesticides are known animal carcinogens or teratogens. DDT, dieldrin, and other chlorinated hydrocarbons are known to be accumulated in human tissue and milk. High levels have been associated with hypertension, arteriosclerotic cardiovascular disease, and possible diabetes (Moses, 1983).[1]

Widespread pesticide use is stimulating the development of resistant insects and other pests, thereby reducing the effectiveness of current pesticides. Significant levels of resistance can develop within twenty generations (McEwen, 1978). Similarly, the use of sub-therapeutic doses of antibiotics on crops and food animals is hastening the development and transmittal to man of bacteria that are resistant to the antibiotics used to control disease (Levy, 1984).

The application of modern pesticides is only one source of food-borne risk. Bacterial contamination of improperly preserved foods (by botulinum, salmonella, or listeria monocytogenes, for example) can lead to severe illness or death (Tanya Roberts, 1985). Listeria monocytogenes were implicated in the death of tens of Los Angeles residents from consuming Jalisco brand cheeses in autumn 1985. In addition, dietary plants contain an impressive array of naturally occurring mutagens and possible carcinogens and anticarcinogens (Ames, 1983; Ames et al., 1987; Hambraeus, 1982). Some of these toxins are produced by molds that may contaminate fruits, grains, nuts, and derived products. Two such mold-produced toxins, aflatoxin and sterigmatocystin, are potent carcinogens. Other mutagens and possible carcinogens occur naturally in fruits and vegetables or are produced by cooking or charring foods (especially meat). Average human intake of these naturally occurring toxins is likely to be on the order of several grams per day, perhaps 10,000 times higher than dietary intake of residual synthetic pesticides (Ames, 1983). The ubiquity of these substances suggests that carcinogenesis depends in a complex way on the interaction among these and other substances.

Plant breeders may attempt to increase the levels of naturally occurring toxins to promote pest resistance. In at least one case, this effort was too successful: The resulting potato cultivar could not be used because the level of glycoalkaloids it produced was believed to be toxic to humans. (Two of the glycoalkaloids in potatoes, solanine and chaconine, are strong cholinesterase inhibitors and possible teratogens; Ames, 1983.) Although this issue has apparently not been addressed, it seems plausible that at least some of the varieties of produce that are grown organically contain higher levels of these naturally occurring toxins.

ORGANICALLY GROWN PRODUCE

Organically grown fruits and vegetables are produced without the application of pesticides or other synthetic chemicals. In California, any food that is labeled as organic must display a label stating that it was produced in accordance with the applicable section of the Health and Safety Code (Sec. 26569.11). This section requires that organic food be produced, harvested, distributed, stored, and packaged without applying any pesticide, synthetic fertilizer, or growth regulator. Moreover, growers are not allowed to apply any such chemicals to their fields for a year before planting annual or biannual crops, or for a year before the emergence of buds on perennial crops. Finally, regardless of how the produce was grown, if it contains more than 10 percent of the legal tolerance for any pesticide it cannot be sold as organic.[2]

Inspection and other enforcement activities are limited. Surely, there are cases of fraudulent sales, and there is little even a conscientious grower can do to prevent his produce from absorbing the environmentally persistent pesticides that remain in the soil or in irrigation water or are deposited as overspray from neighboring fields. As described below, consumers are aware of these problems and suspicious of the purity of foods sold as organically grown. However, many stores that sell organic produce seem careful to ensure its quality, for example by establishing relationships with a few suppliers on whom they can depend. It is not uncommon for a store manager unsure about the quality of a shipment to place a sign above it stating that the food was represented as organic but that the manager has no other proof.

Some organizations certify food as organically grown; e.g., the primary certification organization in California is the California Certified Organic Farmers (CCOF). CCOF inspects farmers' fields and records and tests their soil annually (private communication from Warren Weber, CCOF president). Other organizations set standards for organic produce, including the International Federation of Organic Agriculture Movements, headquartered in Europe, and the North American Organic Food Production Association (Gail Roberts, 1985).

Some produce is labelled "unsprayed." Use of this term is not formally regulated. Unsprayed items are not sprayed with pesticides, although chemical fertilizers may have been applied. Their qualities should be intermediate to organic and conventional produce.

Organically grown produce constitutes a small but growing share of the fresh produce market and organic-food consumers are a small, unrepresentative sample of the general population (Hall et al., 1989; Franco, 1989). An estimated five percent of American households spends additional amounts to purchase specialty foods that it considers more wholesome, nutritious, or better tasting than conventional supermarket products (Burros, 1986). Organic-food buyers constitute only a small part of this five percent. In the West Los Angeles/Santa Monica area (the site of the study reported here), only a few stores consistently offer a large selection of organic produce and other groceries in contrast to the tens of supermarkets and perhaps hundreds of smaller grocery stores.

The organic-food markets do not cater exclusively to consumers who purchase all their groceries there. Many customers purchase most of their groceries at a conventional supermarket and their produce and other specialty items at the smaller stores. Overall, 43 percent of American consumers regularly purchase groceries from more than one store. Among this general population, prices and location are by far the most frequently cited reasons for selecting a store (Food Marketing Institute, 1986).

Systematic information on the comparative demographic and other characteristics of organic- and conventional-food purchasers is apparently unavailable. Some authors suggest that organic-food purchasers are wealthier and better educated than average (Burros, 1986; Marshall, 1974); organic-food buyers in the focus groups discussed below describe themselves as better informed.

A THEORETICAL MODEL OF CHOICE

The individual's problem of choosing an optimal diet can be conceived as a life-long dynamic program. Choices are based on his or her highly imperfect, changing knowledge about the possible health risks of different foods, the variable but potentially long latency between consumption and ill effects (if any), differences between the effects and possible treatments for acute and chronic illnesses, and uncertainty about how long one will live to eat and enjoy. A model that accounted for all these complexities would be impossible to estimate with available data. Consequently, I present a simplified model that serves as the basis for the empirical analysis described in subsequent sections.[3] Like most economic models of behavior, this model assumes well-informed consumers. As discussed later in this chapter, consumers have limited information about food-borne risk. Nevertheless, the model provides useful insights.

Decisions on food purchases are often made for an entire household. Parents usually make the purchasing decisions that affect not only their own but their children's welfare. I do not explicitly consider the issues involved in aggregating individual utility functions into a household decision rule but assume that household utility functions exist that are analogous to individual utility functions. Although I use the terminology of individual decisions, it should be understood that much of the behavior described applies to households.

Assume that the consumer attempts to maximize his utility subject to a budget constraint. Assume further that his utility function is structured so that the foods he consumes constitute a group that is separable from all other goods. Then attention can be restricted to maximization of the subutility function for foods (Deaton and Muellbauer, 1980). I further assume that this subutility function is additively separable in the quantities of foods and the risk imposed by each:

(11) $U = \Sigma_i [t_i(q_i) + \tau_i(q_i) X_i - r_i q_i (1 - X_i)]$

where q_i is the quantity of the ith food consumed, $t_i(.)$ is a function describing the individual's subjective taste for the ith food, $\tau_i(.)$ represents any incremental difference in taste for the organic version of the ith food, X_i is an indicator variable equal to one if the consumer chooses the organic form of food i, zero otherwise, and r_i represents the incremental risk from consuming the conventional version of the ith food instead of the organic type. The assumption of additivity is very strong: It rules out complementarities among foods and synergies and antagonisms among risks and nutrients. As a first approximation, however, it should be acceptable.[4]

The functions $\{t_i(.)\}$ describe the consumer's subjective evaluation of all aspects of the ith food except the health risk from pesticide residues. They describe his reaction to the taste, appearance, consistency, size, and uniformity of the item, as well as any health risks common to organic and conventional varieties such as risks due to naturally occurring carcinogens or benefits from fiber, vitamins or other nutrients. For foods the individual enjoys, $t_i(q_i)$ is positive and increasing in quantity, at least through some satiation level q_i^*. I assume that $t_i(q_i)$ is concave over all q_i for which $t_i'(q_i) > 0$. For distasteful foods, $t_i(q_i)$ is less than zero for all nonzero quantities.

The functions $\{\tau_i(q_i)\}$ account for any differences between organically and conventionally grown produce other than the levels of residual pesticides. They account for any differences in taste, appearance, texture, consistency, and other qualities. These other qualities include the possible satisfaction one derives from feeling that he or she is exerting some measure of control over food-borne risks, or is behaving in an ecologically responsible manner by choosing to consume organically grown produce. These functions are assumed to be continuously differentiable (except possibly at $q_i = 0$). As a special case, taste and other differences between varieties are assumed negligible, so the $\{\tau_i(.)\}$ are identically zero.

The incremental risk parameter r_i describes the risk from eating the conventionally grown version of the ith food instead of the organic version. If organic foods pose offsetting risks, r_i could be negative. In principle, the risk includes changes in the probability of living to any age and changes in the probabilities of suffering various nonfatal illnesses due to ingesting pesticides. The parameter r_i is intended to summarize all of the health risks from pesticide residues, ranging from acute but mild food poisoning through latent cancer, sterility, neurotoxic effects, and birth defects. This specification assumes the risks can be linearly aggregated and that total risk is proportional to quantities of foods. While linearity of response is an acceptable approximation to many dose-response models over the limited range of doses usually ingested, it does rule out threshold models. Moreover, it implicitly assumes reciprocity of dose: The effect is proportional to the total quantity ingested, regardless of how the dose is distributed over time. Any nonlinear pharmacokinetic effects, due, for example, to the ability of the liver or other organs to detoxify or eliminate residues from the body, are ignored.

The consumer seeks to maximize the subutility function (11) subject to a budget constraint

$$(12) \qquad \Sigma_i \, (\pi_i \, q_i + p_i \, q_i \, X_i) \leq B$$

where π_i is the price of the conventionally grown version of food i, p_i is the price premium for the organic version (the difference between the organic and conventional prices), and B is the consumer's food budget. The consumer sets the food budget B by balancing the marginal utility he can obtain from greater expenditure on foods against the marginal utility from additional spending in other categories, perhaps in the first stage of a two-stage budgeting algorithm.

The solution to the consumer's diet problem may be characterized by forming the Lagrangian and differentiating to obtain the Kuhn-Tucker conditions:[5]

(13) $$\frac{t_i' + \tau_i' X_i - r_i (1 - X_i)}{\pi_i + p_i X_i} \leq \lambda \text{ (equality must hold if } q_i > 0)$$

(14) $$\begin{cases} \dfrac{p_i}{\tau_i + r_i} \geq \lambda \text{ if } q_i > 0 \text{ and } X_i = 0 \\[2em] \dfrac{p_i}{\tau_i + r_i} \leq \lambda \text{ if } q_i > 0 \text{ and } X_i = 1 \end{cases}$$

(15) $\quad q_i \geq 0$

(16) $\quad X_i (1 - X_i) = 0$

(17) $\quad \Sigma (\pi_i q_i + p_i q_i X_i) \leq B.$

These conditions may be easily interpreted. The first determines the allocation across foods, by equating the ratio of the marginal taste minus marginal risk to the unit price. If the marginal benefit (taste minus risk) is high for some food, relative to its price, the consumer can gain by consuming more of that food and less of others.

The second condition determines the choice of organic versus conventional versions of each food consumed. Consider first the simple case in which, except for pesticide risks, the consumer is equally well satisfied by organic or conventional versions. Then $\tau_i \equiv 0$ and (14) simplifies to

(18) $$\begin{cases} \dfrac{p_i}{r_i} \geq \lambda \text{ if } q_i > 0 \text{ and } X_i = 0, \\[2em] \dfrac{p_i}{r_i} \leq \lambda \text{ if } q_i > 0 \text{ and } X_i = 1. \end{cases}$$

Imagine that the alternative foods are ranked in ascending order of the incremental premium/risk ratio p_i/r_i (i.e., the price of incremental risk reductions). The consumer will choose the organic version of all foods he eats before a certain point on the list, and the conventional version of all foods beyond that point. The dividing point is where the premium/risk ratio is equal to the critical value λ^*. This value is the consumer's WTP to avoid pesticide risks and is determined through condition (13) and the budget constraint (17). Moreover, since λ^* appears in both (13) and (14), the consumer will efficiently minimize risk by choosing between substitution of organic for conventional foods and reduced consumption of other risky foods, organic or conventional.

Since tastes and budgets vary across consumers, each may have a different WTP. However, if all consumers have the same information about the incremental risk r_i, they will rank the foods in the same order and the pattern of choice between organic and conventional produce will be consistent across individuals. Consequently, the ratio of the quantity of organic to conventional foods sold will be a monotonically decreasing function of the incremental premium/risk ratio. Everyone who buys any organic food, who also buys the food with the lowest premium/risk ratio, will buy the organic version. For foods with higher premium/risk ratios, the proportion of buyers who find it worthwhile to purchase the organic type will progressively fall until for some foods there may be very few consumers that buy the organic type. Similarly, if the premium changes over time some consumers will switch from one type of produce to the other.

If consumers are not indifferent between the taste and appearance of organic and conventional foods this consistency may be broken. If consumers derive satisfaction from the perceived ecological or political ramifications of buying organically grown produce, their threshold λ^* may increase, but as long as this satisfaction does not vary with the particular item or quantity (that is, if $\tau_i(q_i) \equiv \tau$ for all i such that $q_i > 0$) the consistency criterion will hold. However, if the consumer's preference for organic produce varies by item, condition (14) applies. In this case, consumers order foods by the ratio of the price premium to the sum of the incremental risk and taste and buy the organic form of the foods whose risk reduction plus (or minus) incremental taste are worth the premium. Hence, consumers can quite rationally purchase the conventional form of a food with a low premium/risk ratio and the organic form of one with a high premium/risk ratio, if the differences in flavor and appearance outweigh the risk difference.

Note that the choice between organic and conventional varieties is independent of the quantity of each food one consumes.[6] If the premium/risk ratio is smaller than one's WTP, one should purchase the organic variety of the food whether it constitutes a large or small part of the diet.

This model assumes that consumers have detailed information on the relative hazards of different foods. As described in Chapter 15, however, there is great uncertainty about the risk posed by alternative foods and pesticides. Consumers are likely to group foods into broad risk categories, so that r_i will be the same for all foods within a category. However, consumers should still choose between organic and conventional varieties using the incremental premium/risk ratio.

The subsequent chapters estimate the premium/risk ratios for various produce types. The consumer's choice between organically and conventionally grown items implicitly reveals whether his or her WTP to reduce pesticide risks is greater or smaller than the price of the risk reductions available by purchasing the organic type of each food. By estimating the premium/risk ratios it is possible to infer how large or small an individual's WTP must be to be consistent with his or her choices among foods.

CONSUMER FOCUS GROUPS

A series of focus-group interviews was conducted to learn more about how consumers choose between organically and conventionally grown produce. The contingent-valuation portion of this study relies on a survey of the focus-group members.

Focus groups are widely used in marketing research and are beginning to be employed for environmental-benefit assessment (Wells, 1974; Payne, 1976; Smith et al., 1985; Desvousges, 1986; Mitchell and Carson, 1986). As contrasted with conventional surveys, focus groups can be better for exploratory research, when the researcher seeks to generate hypotheses before knowing precisely what questions to ask in a survey. Focus groups are often used when developing instruments for a larger survey.

A focus group typically consists of 8 to 15 participants who discuss their views on the topic of interest for about one and one-half hours. The discussion is lead by a moderator, whose goal is to keep the discussion focused on the relevant topics and to ensure that all participants contribute. Because the group discussion is somewhat unstructured it is difficult to accurately summarize the opinions and their distribution within a group. One can informally survey the participants, as done here, but the responses may be influenced by any discussion that has taken place before respondents answer the survey questions and the sample is small and may not be representative of the population of interest. Consequently, the statistical results are only suggestive.

Four focus groups were conducted. All of the participants were primary grocery shoppers for their households. Participants in two of the groups regularly shopped at stores selling organically grown produce and frequently purchased organically grown items. Many shopped at the stores from which the price data were obtained. Participants in the other groups usually shopped at conventional supermarkets and rarely or never purchased organically grown items. Again, many of these participants shopped at the supermarkets from which prices were collected.

The participants were recruited by a firm that specializes in organizing and conducting focus groups. Most were drawn from a list of past or recommended participants maintained by this organization although some of the organic-food purchasers were recruited at the markets. All participants received a cash payment of $30 at the end of their session. Twelve individuals were recruited for each group: Because some failed to arrive, the total number of participants was 22 in the organic-food buying groups and 23 in the conventional-food buying groups.

The sessions began with discussion of the types of stores from which participants usually purchased produce, followed by discussion of recent, well publicized, local incidents involving food-borne risks: the ban on watermelon sales in July 1985 and the death of several infants from contaminated Jalisco brand cheeses that autumn. This introduction was intended to help the participants start thinking about food-borne risks. Afterwards, the discussion concerned the extent of the participants' concern about food-borne risks, their knowledge of organically grown produce and the differences in price, risk, taste, appearance, and nutrition between it and conventionally grown produce, the comparative magnitude of the risks from conventional produce and other common risks, and participants' management of these other risks.

At the end of the sessions participants completed a brief questionnaire, Responses to questions about the perceived numerical risks from conventionally and organically grown produce, and WTP for whichever type the participant preferred, provide the basis for the contingent-valuation estimates described in following chapters. The participants were asked to estimate the lifetime fatality risk from cancer or other adverse effects of eating organically and conventionally grown produce. To help in answering these questions, the estimated annual probabilities of death from a wide range of risks, including various types of cancer, accidents, occupational and recreational activities, were illustrated using an attached risk ladder (Fig. 23) Respondents were asked to mark their risk ladders to indicate their subjective estimates of the risk of "eventually dying from cancer or other disease caused by the pesticides and other residues and toxins" on conventionally or organically grown produce, based on their own consumption patterns. Because the risk ladder forces respondents to compare their estimates of the risk from eating produce to the estimated risks of many other causes of death, it should enable them to translate their qualitative beliefs into more accurate numerical estimates.[7]

After estimating the risk of consuming each type of produce, respondents were asked the maximum additional percentage they would be willing to pay to purchase the type of food they believe to be less risky (usually organically grown) instead of the other type. To discover any differences in perceived risk between different types of produce respondents were asked to list specific items for which they thought the organic version was safer, riskier, or about as safe as the conventional version. Respondents were also asked about their management of other risks: whether they smoke cigarettes, use automobile seat belts, eat red meat, and drink tap water.

CONSUMER PERCEPTIONS OF FOOD-BORNE RISKS

One of the major reasons for conducting the focus groups was to understand the reasons some individuals are willing to pay substantial premiums for organically grown produce. The major reason the organic-food buying participants gave was to protect their own and their family's health. They were fearful of ill effects due to pesticide residues and to other applied chemicals such as growth stimulators and fertilizers. They suggested that these effects might include cancer, sterility, allergies, digestive problems, asthma, premature senility, general loss of energy, loss of hair, and an entire range of insidious but unknown effects. The organic-food buyers claimed that many of these risks were cumulative and that adverse effects might take many years to become evident. Some claim that eating organically grown produce will not only prolong life but also enhance current health. The participants suggested that, since the organic growing process involves fewer "insults" to the food, organically grown produce is more nutritious. Specifically, respondents claimed the use of artificial chemicals in growing produce could destroy the nutrients in the food, or at least prevent their assimilation.

Both types of consumers listed price and convenience as important factors in choosing produce. The organic-food buyers were of course more likely to think that the reduction in risk was worth the increased unit price. Several suggested that the net difference in cost is not as great as one might think, since by eating organically grown produce one is supposedly less likely to purchase comparatively expensive junk foods and meat and will save on medical expenses. Conventional-food buyers believe organically grown produce to be substantially more expensive than conventional produce. They also found other aspects of convenience in addition to the limited availability of organically grown produce. They claim that organically grown produce does not keep as long, and note that one cannot eat out very often if he or she wishes to avoid consuming conventionally grown produce.

An additional reason for purchasing organically grown produce might be described as one of ecological or political consciousness. A few of the organic-food buyers made statements about "owing it to the planet" to reduce the "wear and tear" caused by environmentally unsound agricultural practices.

Some organic-food buyers expressed strong opinions about differences in taste between organically and conventionally grown produce. However, it appears unlikely that many of them could distinguish between organically and conventionally grown produce in a blind taste test since the perceived difference in taste appears to be associated, in large part, with a sense of "feeling good about yourself" that comes from knowing one is eating carefully. Many participants, in both organic- and conventional-food buying groups, believe there is no significant difference in the taste of most items. For some items, such as tomatoes, the organically grown item is sometimes vine ripened and consequently juicier and tastier. However, some organic-food buyers report that vine ripened, chemically treated tomatoes are just as tasty.

Limited empirical evidence suggests that systematic differences in taste are negligible. Schutz and Lorenz (1976) conducted a blind taste test using specially grown samples of lettuce, broccoli, green beans, and carrots. The samples were carefully grown under typical conventional or organic conditions. In a blind test, the tasters (college students) indicated no significant differences, except they preferred the conventional to the organically grown carrots. In a second test the types of produce were labelled (correctly for some tasters and incorrectly for others). The only significant differences were that the tasters preferred the organically grown broccoli, and they preferred whichever product was labelled organic.

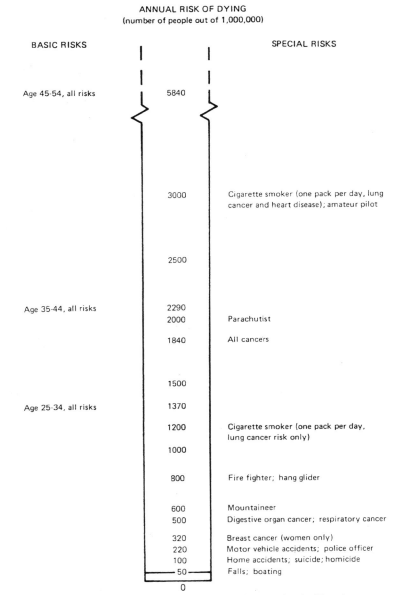

Fig. 23 — Risk ladder showing annual mortality probabilities associated with various causes, general U.S. population or relevant subset.

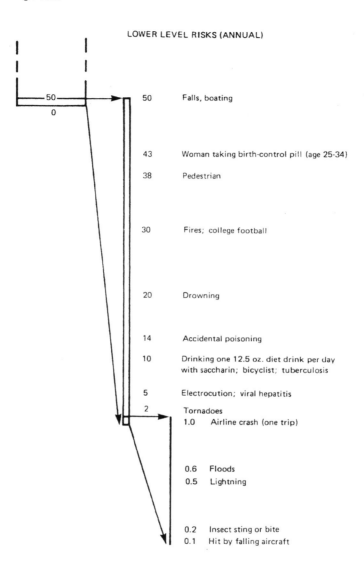

LOWER LEVEL RISKS (ANNUAL)

50 — Falls, boating

43 — Woman taking birth-control pill (age 25-34)

38 — Pedestrian

30 — Fires; college football

20 — Drowning

14 — Accidental poisoning

10 — Drinking one 12.5 oz. diet drink per day with saccharin; bicyclist; tuberculosis

5 — Electrocution; viral hepatitis

2 — Tornadoes

1.0 — Airline crash (one trip)

0.6 — Floods

0.5 — Lightning

0.2 — Insect sting or bite

0.1 — Hit by falling aircraft

Fig. 23 (continued).

The focus-group participants perceive a greater difference between the appearance of organically and conventionally grown produce. Most think that conventional produce is cosmetically very attractive, but organic-food purchasers think it sometimes looks artificial. Most of the conventional-food buyers are aware of the use of wax and colorants on conventional produce and many claim that they would pay more for produce that was not so treated, although some object to the idea that they should pay more to avoid additives. Conventional-food buyers object to finding insects on their food but organic-food buyers are willing to overlook this disadvantage and argue that one should wash produce anyway, to remove soil as well as insects.

Conventional-food buyers generally think that pesticides and other chemicals and growing practices used in large-scale conventional agriculture do not present a significant health hazard. They appear to base this conclusion on the belief that, if such practices did pose a threat, either the government or the supermarkets would warn and protect them or they would have observed evidence of harm. Many of these participants said they shop only at "high-quality" markets (by which they mean large supermarket chains) and avoid both small groceries and discount stores. Nevertheless, most of the conventional-food buyers seem to think that organically grown produce is a good idea, if one could be sure that it was really organically grown, and if standards of cleanliness and so forth were adequate. Organic-food buyers are also suspicious of whether food sold as organically grown really is, but claim that "you have to give it your best shot" (that is, choose the produce that is claimed to be organic, trusting that, even if it is not as represented, it is at least not as bad as conventional produce). The conventional-food buyers suggested an offsetting risk to organic produce: Since insects may carry disease, the use of pesticides may contribute to food safety.

Organic-food buyers show a sustained commitment. Most of the participants in the focus groups claimed to have purchased organically grown produce regularly for five or more years and recognize that doing so requires a greater commitment of time and money. Some had been unable to purchase organically grown produce when they had lived in other parts of the country because of its limited availability.

A few of the conventional-food buyers may have difficulty visualizing a low-level, cumulative risk. For example, one commented that pesticides need not be dangerous to humans since, when one sprays home insecticide, "it kills the bugs but not the people."[8] In contrast, the organic-food buyers are very worried about such risks: In addition to their concern about food-borne risks, some noted that they are also concerned about using many household cleansers.

Neither organic- nor conventional-food buyers distinguish much among risk differences across foods. There was some suggestion that pesticides were not as serious a threat on items where one peels the skin before eating it (e.g., bananas, oranges, and avocados) or on items that are cooked (e.g., potatoes). However, both types of participants believe that the pesticides are absorbed into the fruit; the recent contaminated watermelon incident is clear evidence of this. One section of the questionnaire asked participants to distinguish between foods for which the risk between organically and conventionally grown items differed. A few noted the difference between peeled and/or cooked and other items, but most thought the difference in risk was consistent across all items.

Participants in both sets of groups made some interesting observations about the differences between organic- and conventional-food buyers. The organic-food buyers characterized themselves as "mavericks"—better informed and more health-conscious than most, although one admitted that organic-food buyers sometimes "make a god out of eating right" and ignore more important risks. They think conventional-food buyers are probably too apathetic or think themselves to be too busy to take responsibility for this important area of their lives. Conventional-food buyers concurred. They claim organic-food buyers are often "fanatics" who are so obsessed with relatively modest food-borne risks that they ignore other risks such as motor-vehicle accidents. They believe that some organic-food purchasers are people who have suffered a bad health scare such as a heart attack or cancer. Conventional-food purchasers describe themselves as price conscious, too busy to suffer the inconvenience of going to specialty food stores, and concerned about having a wide selection of foods available. They also suggested that organic-food buyers may be more concerned about the future than conventional-food buyers.

CHAPTER 14

WILLINGNESS TO PAY FOR ORGANICALLY GROWN FOODS

The revealed-preference estimates begin with estimates of the premium organic-food buyers pay. These are estimated using prices of organic and conventional foods recorded at several local stores. Except for some information on the size of certain items, data to control for other quality differences between samples, such as differences in color, consistency, taste, and freshness were unavailable. Without such data it is impossible to separate the estimated organic premium into components attributable to the presumed absence of pesticides and to differences in other qualities. Because these data provide virtually no information that can be used to control for differences that might affect the incremental taste function $\tau_i(q_i)$, the function is assumed to be uniformly zero.

PRICE DATA

Prices were observed for organically and conventionally grown produce at five stores in West Los Angeles and Santa Monica. The stores included two food cooperatives, one health-food supermarket, and two conventional supermarkets. Except for the conventional supermarkets, all stores offered both organically and conventionally grown produce, although availability varied by week. The cooperatives are open to nonmembers and generally offer the widest selection of organic foods. Both are medium-sized stores and offer a sufficiently wide selection of products that consumers could purchase all of their groceries at either one. The health-food supermarket is larger, the size of a small supermarket, and also offers a full selection of groceries. It is part of a local chain. The two conventional supermarkets represent two of the major chains in the area. They do not sell any organic produce. The two were chosen because they compete directly with the smaller stores due to their proximity. Shoppers can easily purchase specialty items, like organic fruits, at the smaller stores and the bulk of their groceries at the supermarkets.

The food cooperatives offer discounts of about 15 percent to members. Membership is open to all. One cooperative requires a one-time payment of approximately $35, after which the member must work a few hours or pay an additional $5 per month to receive the member discount. Membership terms at the other cooperative are similar. The stores do not offer end-of-year rebates like other consumer cooperatives. Ignoring the initial fee, it is economical to join if the member will spend more than about $35 a month at the cooperative. Individuals who purchase most of their groceries from the cooperative would thus be likely to join, while those who purchase only specialty items are less likely to join, depending on the other benefits they receive from membership.

Nonmember prices are used in the analysis, as they may be more relevant to the marginal organic produce buyer. A substantial share (estimated by the management as one-third or more) of the customers at the cooperatives are nonmembers. Many of those who divide their purchases between the supermarkets and the cooperatives are likely to be nonmembers. Moreover, although the marginal price to members is the discounted unit price, the full cost they pay for access to organic produce must take account of the additional membership costs. If the cooperative does not strongly subsidize members at the expense of nonmember customers, the full price members pay should be close to the nonmember price.

Prices were recorded at five stores and weekly intervals during a ten-week period beginning with the week of May 27, 1985. Since prices and availability of fresh produce vary seasonally, the results pertain only to the late spring/early summer period. Initially, prices for all types of produce were recorded except items that serve primarily as a garnish or trimming on other foods (e.g., parsley) and relatively unusual items (e.g., English cucumber, pomegranates). After discarding foods for which organic and conventional prices from the same store, or both organic and conventional prices from more than one store, were not obtained, 27 foods remain for analysis. The number of price observations for each food ranges from about 35 to 90, but most have about 50.

REVEALED WILLINGNESS TO PAY

Consistent with the model of consumer choice described in the previous chapter, each store faces demand curves for organically and conventionally grown varieties of each food. These demands are conditional on the spread between the organic and conventional prices. If the spread narrows, some consumers will switch to the organic variety; if it widens, others will switch to conventional. The demands may vary between stores reflecting characteristics such as location, atmosphere, diversity of products, length of check-out lines, and so forth.

Each store can offer organic or conventional produce along a corresponding supply curve. These supply curves are largely determined by the wholesale prices the store can obtain, plus a standard markup.[1] The supply curves shift over time, primarily due to the seasonal characteristics of fresh produce that influence growing and transportation costs.

Premiums for organic produce can be estimated under alternative sets of plausible conditions. If the store-level supply curves are perfectly elastic over the relevant quantities and time it does not matter whether demands shift or not. If the supply curves shift in parallel the organic premiums are constant; if not, the premiums, and the number of consumers buying each type, will vary, but all of the price observations can be used to estimate the premiums at any time during the season. Since each market is small relative to both the local supply and demand, the elastic supply assumption is plausible.

The assumption that the supply curves move in parallel is reasonable over the late spring/early summer period cover by these data, since most of the fresh produce sold then is grown domestically, and seasonal effects are similar for both organic and conventional produce. In contrast, the supply curves are less likely to move in parallel during the winter when a substantial share of conventional produce is imported but nearly all organic produce sold is grown in-state. As a check on this hypothesis, I test whether the price differences for each food remain constant over the ten-week interval.

Alternatively, fixed premiums can be estimated if the demand curves for each store are fixed and the supply curves remain parallel over the ten-week period. These assumptions are tenable, but less attractive than the elastic supply condition. Although store-level demand curves are likely to be sensitive to the prices charged at other local stores, since consumers may compare prices between stores, this factor may not be important if transient price variations between stores are not large enough to induce many consumers to bear the search costs required to capitalize on them. One would not expect regional demands to vary much over a season, except that specific foods are more popular around holidays with which they are traditionally associated and demand for particular foods can be severely depressed by adverse publicity about a temporary and unusual risk.[2] Both of these effects may have operated on watermelons during the data-collection period.[3] However, watermelon is not included in the analysis, and none of the fruits that are appear to be close substitutes or complements, so this incident should not have caused demand curves to shift. Moreover, focus group participants claim the scare did not affect their purchases of foods other than watermelons.

If either set of conditions—perfectly elastic supply or fixed demands coupled with parallel supply curves—hold, price differences between organic and conventional varieties can be estimated. The parameters of the following linear regression model were estimated independently for each food in the sample:

$$y_{ijt} = \pi_i + p_i \, v + \beta_j \, m_j + \gamma_t \, w_t + v' \, \Sigma_\lambda \, \alpha_\lambda \, w_\lambda$$

$$+ \, \delta \, s + \varepsilon_{ijt}$$

where

y_{ijt} = price of food i at market j and week t;
π_i = estimated price of the conventional variety of food i at week 5.5 and market 5;
v = indicator equal to 1 if item is organically grown, 1/2 if unsprayed, and 0 otherwise;
v' = indicator equal to 1 if item is organically grown or unsprayed and 0 otherwise;
m_j = indicator variable equal to 1 if market j, j = 1,2,3,4 (default is market 5);
$w_\lambda = (week - 5.5)^\lambda$, $\lambda = 1,2,3$;
s = size of individual items (+1 if labeled large, -1 if labeled small, 0 otherwise); and
ε_{ijt} = random error with mean 0, distributed independently and identically for all i, j and t.

The coefficient p_i estimates the price premium for organic produce. A small number of items were advertised as "unsprayed." Since these fall somewhere between organically and conventionally grown produce in terms of potential risk, taste, and appearance, the organic produce indicator v is coded as 1/2.

Market and time effects are included to avoid confounding organic premiums with other factors. The market effects potentially reflect each store's choice of suppliers, the demand for its location, and other characteristics. One might expect these effects to be consistent across foods, but the estimates are variable and generally small. Few are statistically significant, indicating small systematic price differences between stores. When the models are estimated without including store effects, most of the estimated premiums do not change significantly. An alternative to this fixed-effect specification would have been to pool the data across foods to obtain more precise estimates of the store effects, assuming these are consistent across foods. However, the store effects may not be consistent and the time effects are not consistent, so this strategy could bias the premium estimates.

To avoid estimating implausible week-to-week variations that could be caused by correlations between time, store, and produce type, a certain amount of structure was imposed on the time effects. *A priori* I chose to represent time trends using a cubic polynomial since this function is sufficiently general to allow for a variety of trends (for example, transition from one fixed price to another) while maintaining reasonable smoothness.

Potential differences in the price trends for organic and conventional produce were tested by introducing interactions between the organic indicator and the time variables. For each food, the model was estimated without interactions, with only a linear interaction, with linear and quadratic interactions, and with linear, quadratic, and cubic interaction terms. For each food the model where the highest level interaction was statistically significant at the five percent level was selected. Of the 27 foods for which premiums are estimated, the hypothesis of a constant premium is not rejected for 12, and these are estimated without interactions. For 10 foods only a linear interaction is included, for three linear and quadratic interactions are included and the regressions for two foods include all three interactions terms.

Sensitivity of the estimated premiums to alternative model specifications was examined. For each food the models were estimated including each of the three levels of interaction between time and organic prices. Each model was also estimated using only data from the stores that carried both varieties of the food. In general, the estimated premiums were very similar across specifications. The only substantial difference was for bell peppers, where the estimated premium was dependent on which time variables were included. This dependence is due to the high correlation between variety and time since organic peppers were available only at the end of the period. Since the interactions between time and variety were never statistically significant, the model without interactions was accepted

The estimated regressions for each food are presented in Table 10. Overall, the regression model describes the price data well. Most of the R^2 values are large and the root mean squared errors small relative to the mean prices. The store effects are usually small and rarely significant, whereas the time trends are sometimes significant over the period.

For most foods, the estimated organic premium is large relative to the price of the conventional item and statistically significant. For a few items (avocados, red and green leaf lettuce) the estimated premium is near zero. Occasionally, it is negative (significantly so for grapefruit). The median ratio of the premium to price of the conventional item is 46 percent; the quartiles are 82 and 6 percent.

At least one of the interactions between time and variety is statistically significant for 15 foods. For these, the hypothesis of a constant premium over the period is rejected. The premium used is the estimated premium at the midpoint of the ten-week study period.

REPORTED WILLINGNESS TO PAY

Focus-group were asked participants to report their incremental willingness to pay (as a percentage of their current spending on fruits and vegetables) for the type of produce they believe to be safer (almost always the organic type). Table 11 summarizes the participants' responses, stratified by the type of produce usually purchased.[4]

The conventional-food buyers reported low to moderate willingness to pay. None of the 23 respondents would be willing to pay more than an additional 20 percent to obtain their preferred type of produce (all three responses in the 20 to 39 percent range are 20 percent). Six (26 percent) of the conventional-food buyers report that they would not pay more to buy either type of produce, that is, they would always buy the less expensive type. Five (22 percent) believe that conventionally grown produce is safer than organic, but none would pay more than 10 percent more to purchase conventionally instead of organically grown (the table shows these five as having negative WTP for organic produce). Twelve of the respondents, just over half, believe that organically grown produce is safer and would be willing to pay more to purchase it. However, three quarters of these (9) would not pay as much as an additional 20 percent, and the remaining three would pay only an additional 20 percent of their typical produce bill.

Table 10

ORGANIC PREMIUM REGRESSION MODELS

Independent variable		Broccoli	Celery	Spinach	Avocado	Carrot	Cucumber	Pepper
Mean commercial price		57.909 (9.67)	53.070 (11.82)	58.410 (20.18)	122.87 (9.39)	29.897 (18.97)	39.704 (6.71)	90.196 (20.04)
Organic premium		62.507 (8.00)	28.624 (5.49)	22.147 (6.33)	-2.240 (-0.08)	24.500 (13.11)	14.915 (1.87)	43.358 (4.59)
Market indicators (default = 5)	1	13.210 (1.88)	-2.593 (-0.44)	10.863 (3.04)	-24.760 (-0.76)	-0.607 (-0.32)	13.256 (1.96)	-2.800 (-0.50)
	2	-11.719 (-1.33)	-12.476 (-1.89)	15.625 (3.42)	34.540 (1.05)	-8.207 (-3.06)	-1.735 (-0.19)	3.293 (0.53)
	3	11.365 (1.53)	-10.956 (-1.68)	-9.035 (-1.87)	6.668 (0.30)	-6.937 (-3.39)	-2.979 (-0.40)	-27.513 (-4.46)
	4	4.319 (0.53)	-16.598 (-2.67)	0.140 (0.03)	-10.511 (-0.58)	-10.055 (-4.50)	1.487 (0.19)	-6.448 (-1.03)
Time trend (in weeks, centered at week 5.5)	Linear	5.724 (2.83)	0.172 (0.10)	6.126 (5.44)	-5.430 (-1.04)	-0.295 (-0.57)	-5.553 (-2.32)	-2.946 (-1.70)
	Quadratic	-0.134 (-0.36)	-0.018 (-0.05)	0.387 (2.34)	-0.348 (-0.44)	-0.096 (-1.17)	-0.285 (-0.78)	0.061 (0.23)
	Cubic	-0.176 (-1.42)	0.005 (0.04)	-0.128 (-1.93)	0.747 (2.32)	0.016 (0.47)	0.210 (1.39)	0.027 (0.25)
Time trend/organic interactions	Linear	-1.620 (-0.96)		-3.125 (-3.86)			7.945 (2.09)	
	Quadratic	1.464 (2.45)					2.336 (4.04)	
	Cubic						-0.941 (-3.97)	
Item size			5.000 (0.79)					
Sample size		56	47	50	45	53	55	44
Root mean squared error		16.19	12.59	8.07	35.99	4.18	15.05	12.41
R squared		.869	.631	.853	.423	.912	.807	.695

Table 10 — continued

Independent variable	Red Onion	Spanish Onion	Potato	Yellow Squash	Zucchini	Tomato	Apple
Mean commercial price	32.708 (19.87)	25.598 (21.91)	35.550 (19.32)	46.544 (7.71)	42.052 (8.05)	67.367 (9.12)	78.668 (22.99)
Organic premium	3.909 (1.24)	38.499 (25.74)	2.561 (1.35)	37.425 (3.67)	37.165 (6.39)	116.20 (16.32)	37.124 (8.80)
Market indicators (default = 5) 1	21.636 (11.04)	3.557 (2.29)	9.330 (4.25)	96.101 (4.61)	14.117 (2.20)	-9.945 (-1.10)	-6.335 (-1.41)
2	31.649 (9.54)	-5.243 (-2.47)	13.635 (5.17)	16.752 (1.76)	-16.032 (-1.99)	-23.338 (-2.33)	6.642 (1.26)
3	4.462 (2.06)	5.315 (3.49)	-0.222 (-0.08)	-1.488 (-0.14)	-13.253 (-1.88)	8.159 (0.94)	2.977 (0.61)
4	3.658 (1.63)	-0.065 (-0.04)	-4.605 (-1.76)	10.378 (1.34)	2.043 (0.28)	-3.135 (-0.28)	-11.270 (-2.50)
Time trend (in weeks, centered at week 5.5) Linear	0.571 (0.96)	1.338 (3.33)	-2.033 (-2.97)	-1.652 (-0.67)	-2.470 (-1.38)	0.159 (-0.06)	2.614 (1.89)
Quadratic	0.423 (3.56)	0.042 (0.66)	-0.018 (-0.18)	2.442 (4.88)	0.285 (1.00)	-0.403 (-1.03)	0.252 (1.18)
Cubic	0.007 (0.19)	-0.022 (-0.86)	0.017 (0.42)	-0.367 (-2.37)	-0.015 (-0.13)	-0.138 (-0.89)	-0.117 (-1.36)
Time trend/ organic interactions Linear	-0.931 (-1.69)	-1.208 (-3.65)	1.499 (2.70)	5.355 (1.82)		-11.080 (-5.24)	-6.593 (-4.73)
Quadratic	-0.516 (-2.60)			-3.326 (-3.81)			
Cubic							
Item size		6.500 (2.84)		19.500 (1.83)		13.339 (2.15)	21.016 (9.42)
Sample size	44	55	56	35	55	74	66
Root mean squared error	4.18	3.24	5.16	15.10	14.74	23.84	11.42
R squared	.930	.973	.761	.902	.734	.861	.834

Table 10 — continued

Independent variable	Banana	Bing Cherry	Lemon	Orange	Peach	Apricot	Leaf Lettuce
Mean commercial price	25.595 (12.39)	142.70 (8.52)	75.041 (23.28)	44.248 (14.67)	54.018 (8.73)	59.853 (18.00)	49.683 (36.76)
Organic premium	11.663 (6.66)	94.822 (4.99)	-8.358 (-1.36)	6.845 (0.91)	67.431 (8.25)	14.783 (4.45)	1.379 (0.61)
Market indicators (default = 5) 1	8.875 (3.54)	69.816 (3.44)	-9.542 (-1.30)	0.040 (0.00)	16.704 (2.23)	38.436 (9.36)	29.008 (11.62)
2	10.133 (4.07)	1.563 (0.06)	-11.042 (-1.51)	-15.960 (-1.87)	3.641 (0.36)	36.686 (8.14)	25.848 (10.50)
3	4.849 (1.87)	35.505 (1.65)	-16.596 (-3.10)	-14.484 (-1.96)	3.539 (0.44)	23.713 (5.00)	-7.626 (-4.43)
4	0.105 (0.04)	6.965 (0.30)	30.596 (6.80)	-14.320 (-2.99)	7.818 (0.89)	29.940 (4.76)	-5.055 (-2.93)
Time trend (in weeks, centered at week 5.5) Linear	-0.981 (-1.49)	-8.393 (-1.47)	3.518 (2.92)	2.238 (1.68)	-0.454 (-0.19)	0.450 (0.40)	1.991 (3.11)
Quadratic	0.340 (3.20)	-0.146 (-0.16)	0.225 (1.22)	0.178 (0.95)	1.795 (4.54)	-0.342 (-1.52)	0.160 (1.59)
Cubic	0.017 (0.40)	0.520 (1.46)	-0.069 (-0.91)	0.015 (0.20)	-0.251 (-1.70)	0.086 (1.13)	-0.115 (-2.90)
Time trend/organic interactions Linear				-3.236 (-3.06)	-11.420 (-4.59)		-2.173 (-2.39)
Quadratic							-0.039 (-0.27)
Cubic							0.134 (2.33)
Item size				2.043 (0.57)	9.167 (1.19)	-18.073 (-2.10)	
Sample size	66	52	46	49	57	36	92
Root mean squared error	5.92	45.57	8.97	8.93	18.84	7.61	4.78
R squared	.724	.658	.869	.520	.771	.899	.930

Table 10 — continued

Independent variable		Grapefruit	Green Cabbage	Red Cabbage	Cauliflower	Kiwi	Romaine Lettuce
Mean commercial price		46.805 (17.78)	25.479 (24.24)	38.071 (23.53)	82.348 (16.82)	261.74 (25.21)	47.775 (25.09)
Organic premium		-11.773 (-2.10)	25.000 (19.69)	21.766 (13.09)	-9.526 (-0.59)	16.430 (0.50)	2.215 (0.71)
Market indicators (default = 5)	1	3.829 (0.62)	11.000(a) (17.05)	7.350(a) (7.94)	10.228 (1.54)	13.270 (0.38)	29.785 (7.62)
	2	9.069 (1.50)			6.733 (0.38)	31.970 (-0.91)	29.039 (8.74)
	3	-1.786 (-0.34)	3.985 (2.89)	-9.845 (-4.85)	-2.395 (-0.35)	-20.434 (-1.49)	-5.901 (-2.23)
	4	-10.775 (-2.37)	-6.855 (-4.65)	-18.559 (-8.52)	-10.157 (-1.46)	-77.415 (-5.20)	-9.134 (-3.45)
Time trend (in weeks, centered at week 5.5)	Linear	1.332 (1.09)	0.346 (0.85)	-0.518 (-0.88)	2.495 (1.19)	5.858 (1.50)	1.466 (2.08)
	Quadratic	0.198 (1.18)	-0.058 (-0.92)	0.099 (1.07)	-0.300 (-0.80)	-0.841 (-1.41)	0.149 (1.33)
	Cubic	-0.006 (-0.08)	-0.016 (-0.65)	0.024 (0.66)	0.031 (0.21)	-0.648 (-2.66)	-0.076 (-1.72)
Time trend/organic interactions	Linear	-1.925 (-2.02)	0.701 (2.19)	-1.377 (-2.99)			
	Quadratic						
	Cubic						
Item size		6.155 (1.53)					
Sample size		64	45	44	41	46	46
Root mean squared error		9.37	2.88	4.14	13.94	28.90	5.27
R squared		.261	.966	.956	.330	.753	.935

NOTES: Dependent variable is retail price. All prices are in cents per pound. Item size = 1 if labeled large, -1 if labeled small, 0 otherwise.
(a) Neither market 1 nor 2 sold commercial cabbages during the data-collection period. The coefficient listed for market 1 is for a variable coded 1 for market 1, -1 for market 2. It estimates one-half the amount by which the prices at market 1 exceed those at market 2.

Table 11

REPORTED INCREMENTAL WILLINGNESS TO PAY FOR ORGANIC PRODUCE

Willingness to Pay (Percent of Current Spending)	Organic-Food Buyers		Conventional-Food Buyers	
	Number	Cumulative Percentage	Number	Cumulative Percentage
-10 to -1			5	22
0			6	48
1 to 19			9	87
20 to 39	6	30	3	100
40 to 59	5	55		
60 to 79	1	60		
100 to 119	4	80		
200 to 219	1	85		
Lower quartile	30		0.8	
Median	50		5	
Upper quartile	100		10	

Note: The median and quartiles for the conventional-food buyers are for the absolute value of reported willingness to pay.

The organic-food purchasers report substantially higher WTP, as expected. The median willingness to pay is 50 percent (compared with 5 percent for the conventional-food buyers) and the lowest value (20 percent) is the same as the highest value for the conventional-food buyers.

Reported WTP for organically grown produce is generally consistent with respondents' shopping choices. The distribution of organic-produce buyers' reported WTP is similar to the distribution of estimated organic-premium/conventional-price ratios, but shifted to slightly higher values. The median and quartiles of the reported WTP distribution are 30, 50, and 100 percent, compared with 6, 46, and 82 percent for the estimated premiums. Thus, it appears that many organic-food buyers find some organic-items too expensive, and that if the difference between organic- and conventional-food prices were to rise much, many of the organic-produce buyers might substitute more conventional for organic items. Alternatively, the reported WTP estimates may be "anchored" by the current premiums charged; if the premiums were to increase, organic-food buyers might revise their reports upward.[5]

Similarly, the conventional-food buyers' reported WTP is generally smaller than the observed premiums. The median willingness to pay for this group is only 5 percent, about equal to the lower quartile of the distribution of estimated market premiums. Although many conventional-food purchasers would apparently find it worthwhile to purchase a few organically grown items that have small or negative premiums, the additional costs of traveling to a second market to find these items may easily outweigh the small utility gains.

CHAPTER 15

AVOIDED RISK

Current knowledge of the relationship between human consumption of pesticide residues and health risks is not adequate to reliably characterize the risks. Only a minority of the compounds used have been subjected to state-of-the-art toxicological testing, and even after a compound is tested in bacteria and laboratory mammals, substantial uncertainty remains about its effects on humans. In addition, the identities and quantities of pesticides remaining on produce at time of consumption are not well characterized. As in many areas of environmental and health policy making, substantial outcome uncertainty complicates the analysis.

In this chapter, methods are developed to approximate the relative magnitude of human health risks associated with consumption of pesticide-contaminated produce. Estimates of the risk of one of the most prominent adverse outcomes, cancer, are also developed using dose-response information for a few substances. These estimates are compared with consumers' estimates of the magnitude of the health risks reported in the focus groups.

ADVERSE EFFECTS OF PESTICIDES

Pesticides are poisons, designed to kill selected biological organisms. Unfortunately, it is not possible to precisely target their effects, so it is not surprising that consumption may threaten human health with diverse adverse effects as well. Although the majority have not been thoroughly tested, some pesticides are known to be carcinogens or teratogens in mammals and others are suspected to be. Some pesticides are acutely toxic in miniscule doses while others are comparatively nontoxic.

Evaluating the diverse risks posed by ingesting these compounds requires consideration of the multidimensional nature of health risks. Formally, such risks can be characterized as changes in the probability of suffering specified adverse health effects, including death, at various future times (Cave, 1988). In practice, attention is typically restricted to simple indices or proxies for this constellation of effects, such as the increase in the annual probability of death, or the lifetime chance of contracting a specific disease, such as cancer.

Knowledge of the human health effects of routinely consuming small quantities of pesticides is limited. Indeed, the relationship between nearly all aspects of diet and one of the important and most extensively researched health risks, cancer, is poorly understood (Burkitt, 1982; Hambraeus, 1982; McBean and Speckmann, 1982; National Research Council, 1982, 1987). Because widespread use of modern pesticides developed only in the last 40 years, some of the effects may not yet be apparent. Currently, 40,000 to 50,000 pesticides are marketed in the United States, containing about 1,400 active ingredients registered for use with EPA. Of these, many came into use before modern testing methods, especially for carcinogenicity, mutagenicity, and teratogenicity, were developed and required (Moses, 1983). The National Research Council (1984) estimates that almost two-thirds of pesticides (including inert and active ingredients) have not been subjected to even minimal testing. A survey for EPA found that 24 of the 25 most commonly used pesticides, accounting for about 80 percent of the market, have not been adequately tested (Smith, 1978). Potential synergies among substances are virtually unexplored.

Data on acute toxicity is fairly complete, at least for rats and other laboratory animals (Hayes, 1982; *Farm Chemicals Handbook '85*). Data on chronic effects is extensive for some pesticides, like EDB and DDT, but sparse and of poor quality for others. Many pesticides have not been tested for major chronic effects. According to a Congressional study, of all pesticides sold today, 80 percent have not been sufficiently tested for carcinogenicity, 60 to 70 percent have not been sufficiently tested for teratogenicity, and 30 to 50 percent have not been sufficiently tested for other reproductive effects such as sterility.[1]

Even when a chemical has been extensively tested, most of the test results involve the effects of relatively large doses on laboratory animals, bacteria, and cultured cells. Extrapolation of these results to humans is fraught with uncertainty. Humans live about 35 times longer than laboratory rats, are exposed to a much broader and varied array of potentially hazardous substances, and are genetically much more diverse (Breyer, 1982). Interspecies and low-dose extrapolation remain controversial because they depend on a number of critical assumptions. The most important and uncertain concern selection of the proper form for the dose-response model used to extrapolate from the high doses used on test animals to the low doses typically encountered by humans, pharmacokinetic processes that may lead to nonlinearities in the relationship between the dose administered and the dose that reaches the affected parts of the body (which may depend on the distribution of doses over time), sampling variation in the parameter estimates that leads to wide confidence bands when extrapolating to very low doses, scaling effects between species that differ in tissue distribution and metabolism, and possible differences among humans in sensitivity to toxins. California Department of Health Services (1982) reports the choice of dose-response model alone can affect the estimate of the dose associated with a 10^{-8} cancer risk (from aflatoxin B_1) by a factor of almost one million. Since many of the proposed models fit the observed data equally well and physiological considerations do not strongly support any of the models, it is difficult to know which to choose.[2]

ESTIMATING THE MAGNITUDE OF HEALTH RISKS

In this setting, consumer perceptions of the risks of various foods are likely to be formed primarily by informal aggregation of diverse bits of information. The consumer must decide how much information to gather and evaluate and how to combine it into an overall measure of risk. The information can be combined in many ways, potentially leading to very different behavior. One can attempt to avoid ingesting any substance that is a suspected carcinogen, teratogen, or poison. This approach is a caricature of that embodied in the Delaney Amendment, which requires the Food and Drug Administration (FDA) to ban any food additive found to be carcinogenic. But given the ubiquity of pesticide residues this approach would be almost impossible to implement. Moreover, since one is exposed to hundreds of other toxins through the atmosphere and other sources it would almost surely be more efficient to adopt a less severe standard for food-borne risks and use the resources saved to reduce the risk from other sources (Ames et al., 1987). The extreme opposite approach is a caricature of that traditionally displayed by the tobacco companies in their statements concerning the risks of smoking. Following this approach, the consumer would ignore all information about food-borne risks except when a residue is scientifically proven to be a potent human toxin. Animal tests alone might be insufficient to convince such a consumer that a substance is toxic.

Depending on how much risk information the consumer collects and how he processes it, his estimate of the risk avoided by substituting organically for conventionally grown produce will be more or less accurate. As an attempt to mimic possible methods for aggregating these data, and to assess the sensitivity of conclusions about relative risk to the methods for combining data, I developed a set of "risk indices." These indices use available data on pesticide incidence and toxicity to measure the relative risks between different fruits and vegetables, and between conventionally and organically grown items. They combine estimates of pesticide incidence with a number of indicators of the risk from each pesticide.

In addition to the risk indices, the lifetime human cancer risk from consuming conventionally grown produce is crudely estimated in order to estimate the absolute magnitude of consumer WTP to avoid pesticide risks. This estimate combines the estimated average residual concentrations of pesticides with an approximate estimate of the risk posed by the "average" pesticide. I focus on the risk of cancer, since systematic data on other chronic effects are virtually nonexistent. Because of the uncertainties described above, this estimate may well be in error by one or more orders of magnitude.

The risk indices attempt to incorporate the best available incidence and toxicological information. Incidence information was obtained from summaries of the California Department of Food and Agriculture (CDFA) testing program for 1984. Toxicity information is primarily from the National Institute for Occupational Safety and Health (NIOSH) *Registry of Toxic Effects of Chemical Substances*, which summarizes test results on about 60,000 substances.

The CDFA testing program provides extensive information on the identities and levels of pesticides found on conventional produce; however, comparable information for organically grown food is not available. Nearly all of the food CDFA tests is conventionally grown; any organic produce that may be included is not separately identified. Organic food is assumed to be totally free of pesticide residues. Although organically grown produce is likely to contain some residues, because of polluted air and irrigation water and other factors, as long as the concentrations are substantially smaller than on conventional produce this assumption will be adequate. The risk index measures of the relative risk from eating various conventional foods also measure the incremental risk of eating conventional instead of organic produce.

FREQUENCY AND CONCENTRATION OF PESTICIDE RESIDUES

The California Department of Food and Agriculture annually tests about 7,500 samples of fresh produce drawn primarily from retail and wholesale markets and from supermarket chain distribution centers. For most of the foods analyzed, CDFA tested more than 100 samples during 1984. Summaries of these tests indicate the number of samples of each food tested, the number on which residues were found, and the identity and concentration of each residue. These summaries are assumed to measure the concentrations people ingest when they consume the food, although peeling, washing, and cooking may remove or degrade some residues from the food that is actually consumed.

Using routine multiresidue screening tests, CDFA can detect residues of over 100 pesticides in these samples. Special tests for other residues are sometimes used in addition, but much less frequently. The tests detect pesticide concentrations equal to the legal tolerances or smaller, on the order of parts per million. Residues are detected on about 10 percent of samples tested.

The variances for the estimated average concentrations of pesticides that are detected by the multiresidue screens are quite small. However, the variance for pesticides detected only by special tests is often unacceptably large for constructing risk indices because of the small number of samples tested (often fewer than five). To reduce this variability, I adjust the estimates using a Bayesian approach. I assume a prior distribution for the probability of finding residues and revise it to reflect the test results using Bayes' rule. Specifically, the prior distribution for the likelihood of finding a residue is beta with parameters (θ, γ). The mean estimate of the likelihood is θ/γ, while γ characterizes the precision of the prior. As will be clear shortly, γ can be thought of as equivalent to the number of previous samples on which the prior distribution could be based. If n samples of the food have been tested, of which ρ contained the residue, the posterior distribution is beta with parameters $(\theta + \rho, \gamma + n)$. Thus, the revised estimated probability of finding a residue is $(\theta + \rho)/(\gamma + n)$ and the precision of the estimate has increased from γ to $\gamma + n$. Note that the posterior mean $(\theta + \rho)/(\gamma + n)$ can be expressed as

$$\frac{[\gamma(\theta/\gamma) + n(\rho/n)]}{(\gamma + n)}.$$

That is, the posterior mean is an average of the prior mean and the mean frequency in the sample, weighted by the prior precision γ and the sample size n.

The observed frequencies for each food/pesticide combination were adjusted using a beta prior distribution with parameters (.05, 5). This distribution implies a prior probability of finding a residue of 1 percent. Overall, about 10 percent of samples tested have some residues, but since on the order of ten pesticides are found on each type of produce, the probability of finding a specific residue of one of the pesticides used is on the order of one-tenth. The precision parameter is such that the results of testing five samples are as influential as the prior in determining the posterior mean. For samples of 20 or above the adjusted frequency is nearly identical to the sample frequency, as shown in Table 12.

RISK INDICES

The risk indices span a wide range of reasonable methods for estimating the risk from pesticide residues. Except for the first, all of the risk indices are of the general form:

$$\text{Risk index} = \Sigma_i \, \psi_i \, C_i$$

where the summation is over the pesticides found on a food, the C_i are the expected

Table 12

SAMPLE AND ADJUSTED FREQUENCIES USING A
BETA (.05, 5) PRIOR DISTRIBUTION

Number of Samples, of which One Contains Residues	Sample Frequency	Adjusted Frequency
1	1.00	.18
2	.50	.15
3	.33	.13
4	.25	.12
5	.20	.11
10	.10	.07
20	.05	.04
50	.02	.02
100	.01	.01

concentrations of each pesticide in parts per million, and the ψ_i are weights chosen in accordance with some indicator of the relative toxicity of the pesticides. Note that this formulation is similar to the score function and the cumulative weighted potential ozone depleter emissions used in Part I. In each case, the quantities of a set of similar compounds (measured by concentration, growth or emission rate) are weighted by a factor reflecting their relative hazard (ozone-depletion efficiency or human toxicity) and summed to create an overall index of effect.

The simplest risk index, *Frequency* risk, is the expected number of pesticides found on a food item. It is calculated as the sum over all pesticides of the frequency with which each is found on the food. This index is closest in spirit to the Delaney Amendment caricature described earlier. It assumes that pesticide residues are equally harmful, regardless of the quantity or the specific pesticide. This index might be helpful if it is believed that synergies among pesticide residues (consumed simultaneously) pose risks far greater than equivalent amounts of any single pesticide. A person using this index would seek to minimize his consumption of foods on which multiple residues are found.

The *Concentration* risk index is similar in spirit. It is slightly more sophisticated than the Frequency risk index in that it measures the average quantity of pesticides found on a food, but it treats all pesticides as equally hazardous (that is, $\psi_i = 1$ for all i). Given the limited test data and uncertainty about the effects of different pesticides, counting all pesticides as equally risky may not be unreasonable.

The *Tolerance* risk index measures the residues as a fraction of the tolerance allowed for each food ($\psi_i = 1/$tolerance). It assumes that tolerances are set as if to equalize the risk from each pesticide, that is, inversely proportional to pesticide risk. This index counts a residue of 0.2 parts per million (ppm) for which the tolerance is 1 ppm as hazardous as a 10 ppm residue of another pesticide for which the tolerance is 50 ppm.

Although it is true that tolerances for the more toxic pesticides are generally smaller, tolerance-setting involves other considerations as well, such as the value of the pesticide in controlling pests and so reducing growing costs. Tolerances for the same pesticide vary between foods because they are set no higher than necessary to allow use of sufficient quantities of the pesticide to control the target pests. A tolerance of zero on a food need not indicate that it is unduly hazardous in that application. It may indicate that the pesticide is not useful for controlling the pests that usually infest that crop, or that the manufacturer did not expect sales for use on the crop to justify the cost of submitting data to obtain a tolerance. Tolerances also account for the importance of the food in an average American diet: A food that is eaten infrequently or in small quantities may be allowed a larger tolerance, since human exposure is likely to remain low.

Pesticides are occasionally detected on foods for which the applicable tolerance is zero. In these cases, the value of the Tolerance index would be infinite. To reduce the influence of these few observations, the index is recalculated as though the tolerance for that pesticide and food were 0.05 ppm, one-half the smallest tolerance usually found.

The other risk indices use toxicological data specific to each pesticide. Indices are constructed for both acute and chronic effects and combined to create an overall risk index.

A standard measure of acute toxicity is the LD_{50}. Values were obtained for each pesticide from the *Farm Chemicals Handbook '85* and from the *FDA Surveillance Index*. The LD_{50} is the dose, in mg per kg of animal body weight (or ppm), that kills half of the test animals. Estimated LD_{50}s vary by species and sex of the test animal and by route of administration. The LD_{50}s for oral administration in rats are used, as these are available for all of the pesticides found in the analyzed foods.[3] The *Acute* risk index is simply the sum of pesticide residues weighted by the reciprocals of the corresponding LD_{50}s. To make its value compatible with the other indices, it is multiplied by 100 ($\psi_i = 100/LD_{50}$). Thus the Acute index measures the concentration of pesticides as a percentage of the corresponding LD_{50}s measured in mg per kg body weight.

Chronic toxicity information was obtained from the NIOSH *Registry of Toxic Effects of Chemical Substances* (RTECS, October 1984 edition). RTECS is a database maintained by NIOSH that summarizes toxicity test results for a wide range of health effects on about 60,000 substances. It does not report all test results: It reports only the lowest dose producing the adverse effect among all tests for each health effect, test animal, and route of exposure. Moreover, NIOSH makes no attempt to assess the quality of the tests cited. Consequently, RTECS is likely to overstate the risk of many chemicals since it deliberately reports only the most extreme results. Nevertheless, it appears to be the only concise summary of current information on the diverse chronic effects of pesticides.

Five types of chronic effects are considered: carcinogenicity, neoplasticity, equivocal tumor promotion, mutagenicity, and teratogenicity. Simplistically, RTECS classifies a chemical as carcinogenic if tests indicate that it causes malignant tumors that metastasize (spread) in a mammal. It is neoplastic if it causes tumors that would otherwise be carcinogenic, except that they do not spread. An equivocal tumor agent causes tumors, but it cannot be determined whether they are neoplastic. A mutagenic chemical causes unnatural changes in cells, tested *in vitro*, while a teratogenic chemical causes birth defects. For each effect I consider only whether RTECS reported any tests showing the chemical to produce that effect, regardless of dose, test animal, and route of administration. It is important to recognize that the absence of a report for a particular effect and chemical does not necessarily mean the chemical does not have that effect—the appropriate test may not have been performed.[4] Moreover, the reported test data by necessity reflect the research priorities of recent decades: There is extensive information on carcinogenicity but little on other toxic effects.

Two chronic risk indices were developed using the RTECS data. These indices do not consider potency, but only the types of chronic effects reportedly caused by each pesticide. The *Chronic* risk index stems from the observation that the first two effects are logically included in subsequent effects: Carcinogens constitute a subset of neoplastigens, which in turn constitute a subset of equivocal tumor agents. The second and third effects are assumed to be important primarily because they indicate that the substance may be carcinogenic. Extending this reasoning, mutagenicity is also assumed to be important primarily as an indicator of potential carcinogenicity.[5] The Chronic risk index counts only the most serious reported effect and assigns decreasing arbitrary weights to the less serious effects, since each is presumably a weaker indication of potential carcinogenicity. Set aside the issue of teratogenicity for the moment. If RTECS reports a pesticide as carcinogenic, information about the other effects is ignored and the pesticide is weighted by a factor $\psi_i = 1.0$. If it is not a reported carcinogen but is a neoplastigen, $\psi_i = 1/2$, regardless of other effects. Similarly, a chemical that is not reported carcinogenic or neoplastic but is an equivocal tumor agent is weighted by $\psi_i = 1/4$; if the only reported effect is mutagenicity, $\psi_i = 1/8$. Thus each effect is considered only if no stronger effect is reported, and each is considered a progressively weaker signal of possible carcinogenicity. Teratogenicity is treated as a separate effect, adding 1.0 to the value of ψ_i calculated above if the substance is a reported teratogen.

The alternative chronic risk index, the *Chronic Sum* risk index, weights pesticides according to the number of the five chronic effects considered that are reported in RTECS. The ψ_i are integers ranging from zero to five. This index counts carcinogenicity and mutagenicity as equally undesirable and does not impute one effect from others. It is consequently more sensitive to the number and types of tests that have been performed.

However, it is analogous to the index used by Pedersen et al. (1983) to rank industries and occupation by chemical risk.[6] Note that both chronic risk indices assign a weight of zero to pesticides for which none of the five chronic effects have been reported. As noted, they do not distinguish between pesticides of differing potency.

The *Combined* risk index combines the Acute and Chronic risk measures. Individuals will disagree about the relative values assigned to acute and chronic risks. One method for constructing a combined index is to take a weighted average of the Chronic and Acute risk indices, generating a family of combined indices; a simple average is used here.[7] Since the Acute and Chronic scales are comparable in magnitude (the means across foods differ by less than five percent), the Combined risk index places about equal weight on both kinds of risks.

Table 13 lists the weighting factors ψ_i for the 34 pesticides found on the analyzed foods. It also summarizes the effects documented in RTECS. Table 14 reports the values of the risk indices for the conventional varieties of each of the foods. Since organically grown produce is assumed to be pesticide-free, the values of all of the risk indices are zero for organic items.

ESTIMATED LIFETIME CANCER RISK

Although the lifetime human health risks of consuming the pesticide residues typically found on conventionally grown produce cannot be accurately characterized, rough estimates can be made. Due to data limitations, I do not distinguish between types of cancer or between fatal and nonfatal cancers. This is because the animal experiments on which the following estimates are based are inconsistent in their choice of outcome variables. In some cases, the outcome variable counts any tumor the animal develops; in others, it only counts tumors at a specific site (the liver, for example). The experiments often cannot distinguish between fatal and nonfatal tumors because some animals die of other causes (including sacrifice at the end of the experiment) before their tumors can prove fatal. Summary statistics, such as the TD_{50} described below, consequently do not distinguish between types of cancer, and so my estimate cannot either. The estimates are of the approximate lifetime probability of developing some type of cancer from the pesticide residues found on conventionally grown produce, assuming varied, moderate produce consumption.

Table 13

RISK INDEX WEIGHTING FACTORS AND REPORTED HEALTH EFFECTS OF PESTICIDES

Pesticide	Risk Index Weighting Factors				Reported Health Effects(a)					
	Acute	Chronic	Chronic Sum	Combined	CAR	NEO	ETA	MUT	TER	TD50
Acephate	0.110	0.125	1.0	0.118	0	0	0	1	1	na
Captan	0.011	2.0	5.0	1.006	1	1	1	1	1	100
Carbaryl	0.148	2.0	4.0	1.074	1	0	1	1	1	4
CDEC	0.118	1.0	2.0	0.559	1	0	1	1	0	na
Chlorothalonil	0.010	1.0	3.0	0.505	1	0	0	0	0	1,000
CIPC	0.040	1.5	3.0	0.770	0	0	0	1	1	200
Chlorpyrifos	0.543	1.125	2.0	0.834	0	0	0	1	1	100
Dacthal	0.010	0.0	0.0	0.005	0	0	0	1	0	na
DDT	0.885	2.0	5.0	1.442	1	1	1	1	1	50
Diazinon	0.286	1.125	2.0	0.705	0	0	0	1	1	50
Dicloran	0.025	0.25	2.0	0.137	1	0	1	0	0	20
Dicofol	0.134	1.0	1.0	0.567	1	0	0	0	0	40
Dieldrin	2.000	2.0	5.0	2.000	1	1	1	1	0	1
Dimethoate	0.396	2.0	3.0	1.198	1	1	0	1	1	100
EDB	3.175	2.0	4.0	2.587	1	1	1	1	1	1
Endosulfan	1.429	1.5	4.0	1.464	0	0	0	1	0	1
Ethion	1.042	0.0	0.0	0.521	0	0	0	1	0	na
Glyphosate	0.022	0.0	0.0	0.011	0	0	0	1	1	na
Lindane	0.939	1.5	3.0	1.219	1	1	1	1	0	50
Malathion	0.084	1.125	2.0	0.605	0	0	0	1	1	50
Methamidophos	6.993	0.0	0.0	3.497	0	0	0	1	0	na
Methidathion	2.439	0.125	1.0	1.282	0	0	0	0	1	na
Methomyl	4.878	0.125	1.0	2.502	0	0	0	1	0	na
Methyl bromide	0.029	1.0	2.0	0.514	1	1	0	1	0	100
Methyl parathion	5.882	0.0	0.0	2.941	0	0	0	1	0	na
Mevinphos	4.082	0.0	0.0	2.041	0	0	0	0	0	na
Naled	0.233	0.125	1.0	0.179	0	0	0	1	1	na
Oxamyl	18.519	0.0	0.0	9.259	0	0	0	0	0	na
Parathion	11.765	1.0	4.0	6.382	1	1	0	1	1	3
PCNB	0.006	2.0	4.0	1.003	1	0	0	0	0	100
Permethrin	0.046	0.0	0.0	0.023	1	1	0	0	0	na
Phosalone	0.833	0.0	0.0	0.417	0	0	0	0	1	na
Phosmet	0.432	1.125	2.0	0.778	0	0	0	1	1	na
Vinclozolin	0.010	0.125	1.0	0.067	0	0	0	1	0	na

(a) 1: RTECS reports a positive test for some species.   0: no tests reported positive.
CAR: Carcinogen   NEO: Neoplastigen   ETA: Equivocal tumor agent
MUT: Mutagen   TER: Teratogen
TD50: Estimated daily average dose as a fraction of body weight (in mg/kg) associated with one-half of test animals developing tumors (Gold et al., 1984). (na = not available)

Table 14

RISK INDICES FOR CONVENTIONALLY GROWN PRODUCE

				Risk Index(a)			
Food	Frequency	Concentration	Tolerance	Acute	Chronic	Chronic Sum	Combined
Cucumber	0.964	2.157	42.463	6.761	3.743	8.585	5.252
Leaf lettuce	1.147	1.447	0.256	4.444	1.000	3.091	2.722
Valencia orange	1.369	3.016	0.273	0.454	3.873	8.369	2.164
Spinach	1.161	1.063	3.029	3.878	0.263	1.305	2.071
Potato	0.598	1.924	0.126	0.077	2.402	5.768	1.239
Grapefruit	0.824	1.359	0.193	0.394	1.275	2.648	0.834
Peach	1.379	1.985	0.373	0.285	0.983	4.228	0.634
Green pepper	0.513	0.325	0.182	0.919	0.330	0.750	0.624
Red cabbage	0.289	0.239	0.060	0.755	0.103	0.387	0.429
Green cabbage	0.289	0.239	0.060	0.755	0.103	0.387	0.429
Kiwi	0.378	0.720	0.048	0.239	0.498	1.264	0.369
Apricot	0.175	0.350	0.035	0.052	0.612	1.400	0.332
Celery	0.571	0.681	0.071	0.372	0.286	1.088	0.329
Romaine lettuce	0.340	0.267	0.068	0.529	0.125	0.558	0.327
Broccoli	0.290	0.255	0.054	0.158	0.369	0.772	0.263
Lemon	0.194	0.054	0.025	0.169	0.035	0.067	0.102
Carrot	0.174	0.063	0.082	0.122	0.058	0.189	0.090
Bing cherry	0.209	0.417	0.021	0.010	0.104	0.834	0.057
Tomato	0.274	0.061	0.043	0.031	0.022	0.084	0.027
Apple	0.045	0.048	0.007	0.039	0.003	0.007	0.021
Zucchini	0.158	0.011	0.026	0.015	0.013	0.040	0.014
Yellow squash	0.072	0.011	0.022	0.011	0.004	0.010	0.008
Cauliflower	0.067	0.028	0.030	0.007	0.007	0.032	0.007
Red onion	0.0	0.0	0.0	0.0	0.0	0.0	0.0
Spanish onion	0.0	0.0	0.0	0.0	0.0	0.0	0.0
Banana	0.0	0.0	0.0	0.0	0.0	0.0	0.0
Avocado	0.0	0.0	0.0	0.0	0.0	0.0	0.0
Lower quartile	0.086	0.031	0.022	0.012	0.008	0.033	0.015
Median	0.289	0.255	0.054	0.158	0.104	0.558	0.327
Upper quartile	0.593	1.006	0.173	0.516	0.593	1.384	0.632

(a) Frequency: Expected number of pesticides on an item of food. Concentration: Expected concentration of pesticides in parts per million. Tolerance: Expected concentration of pesticides as a fraction of the legal tolerances. Acute: Expected concentration of pesticides as a percentage of the corresponding LD50s for oral administration in rats. Chronic: Expected concentration of pesticides weighted by an index of reported chronic effects. Chronic Sum: Expected concentration of pesticides weighted by the number of reported chronic effects. Combined: Expected concentration of pesticides weighted by one half the LD50 plus one half the index of reported chronic effects.

The first method uses the TD_{50}s developed by Gold et al. (1984). The TD_{50} is analogous to an LD_{50} for acute risk: It is defined as "the dose rate (in mg/kg body weight-day) that, if administered chronically for a standard period—the 'standard lifespan' of the species—will halve the mortality corrected estimate of the probability of remaining tumorless throughout that period" (Peto et al., 1984). The TD_{50} is essentially the average daily lifetime dose that is associated with the development of tumors in one-half the test animals, adjusted for the frequency of tumors in the control animals. If one assumes the TD_{50}s for humans are the same as those for laboratory animals (primarily rodents) and that the probability of developing cancer is proportional to the average daily dose (over the appropriate interval), an approximate human cancer risk can be estimated.[8]

Gold et al. (1984) do not report TD_{50}s for all of the pesticides detected on the analyzed foods. The values they do report are summarized in Table 13.[9] The median of the reported values is 50 mg/kg body weight-day. This value is assumed to be representative of the average human carcinogenic potency of all the pesticides found on produce items that RTECS reports as carcinogenic.

The approximate average daily human dose can be calculated using the Chronic risk index. The median value of this index, over the 27 foods, is 0.1 ppm. This represents an estimate of the concentration of carcinogenic pesticide residues where pesticides that have been shown to be neoplastic, equivocal tumor agents, or mutagenic are counted as equivalent to smaller quantities of known animal carcinogens. The index assigns additional weight to possible teratogens. Although this biases the estimated average carcinogen concentration upwards, its effect on the estimated risk is negligible compared with other uncertainties.[10]

If a typical individual weighs 70 kg and consumes 1 kg of produce per day, the average dose of carcinogenic pesticide residues is 1.4×10^{-3} mg/kg body weight (= 0.1 x 1/70). If an average daily dose of 50 mg/kg body weight (the median TD_{50}) is associated with a cancer risk of one-half, then, using a linear extrapolation, the estimated cancer risk associated with the average human dose is 1.4×10^{-5} (= 0.5 x (1.4×10^{-3})/50) or about 14 in one million.[11] Assuming the average human lifespan is 70 years, the corresponding average annual risk is 0.2 per million.

A second method relies on the carcinogenic potency measure proposed by Crouch and Wilson (1979). Their measure assumes the dose response model is linear for the region where the probability of cancer is small and uses the "one-hit" model for larger probabilities (the one-hit model, as well as most other commonly used dose-response functions, is approximately linear for small probabilities). They propose using the slope of the dose-response curve as a measure of potency, where the dose is the average daily dose over the animal's lifetime measured in mg/kg body weight-day. Crouch and Wilson report estimates of this potency measure based on epidemiological studies of humans for one of the pesticides found on the analyzed foods: EDB. Their estimate is 0.8 kg-day/mg.[12] If all of the pesticides indicated as carcinogenic in the Chronic risk index are assumed as potent as EDB, the resulting human lifetime risk of cancer is 1.1×10^{-3} (= $1.4 \times 10^{-3} \times 0.8$, the average daily human dose multiplied by the estimated potency). This estimate is almost 100 times higher than the estimate based on the median TD_{50} (1.4×10^{-5}), since EDB is apparently a more potent carcinogen than the other pesticides that have been tested.[13]

A third estimate of the lifetime cancer risk relies on epidemiological studies of the effect of aflatoxin B_1 on human cancer incidence. (Aflatoxin is produced by a mold that grows primarily on corn, peanuts, and grains.) Crouch and Wilson (1982), summarizing several epidemiological studies, report that a lifetime average daily dose of 0.1 μg is associated with a lifetime cancer risk of about 3×10^{-7}. If the median value of the Chronic risk index is used as a representative measure of the concentration of carcinogens in conventionally grown produce, and these carcinogens are assumed to be as potent as aflatoxin B_1, the estimated lifetime cancer risk is estimated as 3×10^{-4}.[14]

The estimates that assume the average dose-weighted potency of carcinogenic pesticides is the same as that of either EDB or aflatoxin are probably too high. As indicated by its TD_{50}, EDB is believed to be a more potent carcinogen than all but a few of the other pesticides for which estimates are reported, and aflatoxin is one of the most potent human carcinogens known. Moreover, aflatoxin's carcinogenic potency is apparently greatly enhanced in individuals exposed to Hepatitis B virus, which is relatively common in the parts of the world where the epidemiological studies have been conducted (Busby and Wogan, 1984). As a result, the estimated potency may exaggerate the potency of aflatoxin alone.

The first estimate, based on the median TD_{50}, is probably the most reasonable of the three, but it too is highly uncertain, for several reasons. The estimated carcinogenic potency of the known animal carcinogens varies over one million fold in test animals. Any attempt to estimate the "average" potency in humans is highly uncertain. Adequate testing to estimate TD_{50}s for many of the pesticides found on these produce items has not been undertaken. If the substances that have been tested are generally more potent than those that have not, the estimate will be biased upwards. Whether this is likely is unclear, however. The likelihood that a pesticide has been adequately tested is primarily related to how recently it has been introduced, not its potential toxicity, as current testing requirements are much stricter than earlier requirements. In addition, there is substantial uncertainty in estimating the TD_{50}s (see Gold et al., 1984), and even in deciding which substances are carcinogens (note the inconsistency in Table 13 between the pesticides considered to be carcinogenic, based on RTECS data, and those for which Gold et al. report TD_{50}s).

The extrapolation from effects of high doses on laboratory animals to the effects of low doses on humans is highly uncertain, for all of the reasons summarized at the beginning of this section. The linear extrapolation used is likely to overstate the risk at low doses. The interspecies extrapolation, assuming the same dose measured in mg/kg body weight-day produces the same effect in each species, produces a lower estimate than an extrapolation that measures dose relative to body surface area. Little is known about the merits of the two approaches: FDA uses the first and EPA the second method (U.S. General Accounting Office, 1987). Possible synergistic or antagonistic effects of combinations of pesticides, or of pesticides and other risk factors, are unknown. Finally, even the estimates of average residue concentrations are uncertain, as they are subject to sampling variability and changes in pesticide use over time and between regions.

An approximate upper bound on the risk of human cancer from pesticide residues can be derived from Doll and Peto's (1981) "guesstimate" that about 35 percent of cancer incidence in the U.S. population could be prevented by changes in diet. This "guesstimate" includes many routes through which diet could affect cancer incidence in addition to direct ingestion of carcinogenic pesticide residues; thus, it is probably a very high upper bound. Among the other routes they include: ingestion of naturally occurring carcinogens or those produced through cooking or by bacterial or fungal growth; alteration of the conditions affecting the formation, transport, activation, or deactivation of carcinogens within the body, as by providing substrates for carcinogen formation (such as nitrites or nitrates); alteration of the intake and excretion of cholesterol and bile acids or the bacterial flora of the bowel; deactivation of short-lived intracellular species (by use of selenium, vitamin E, or β-carotene); alteration of conditions affecting promotion of carcinogenic cells, as through vitamin A deficiency; and overnutrition.

The annual U.S. cancer death rate is about 1,840 per million,[15] 35 percent of which is about 640 per million. Since the average American lives about 70 years, the lifetime risk is about 70 times as great or 45 per thousand. This estimate is about 40 times greater than the largest of the three estimates of pesticide risks derived above (1.1 x 10^{-3}), based on the Crouch and Wilson (1979) estimate of the carcinogenic potency of EDB. Thus, it provides little guidance.

In summary, estimation of the human cancer risk from lifetime consumption of conventionally grown produce is extremely difficult and uncertain. The best guess developed here is that the risk is on the order of 10 in one million. However, this estimate may be incorrect by perhaps two or more orders of magnitude. A better estimate could be made by combining improved estimates of exposure with carcinogenic potency factors for each pesticide. Even so, substantial uncertainty due to missing data and other factors would remain. Furthermore, pesticides may induce other adverse health consequences, including neurotoxic effects and teratogenicity, the risks of which are even less certain.

PERCEIVED RISKS

Focus-group participants were requested to estimate the approximate "risk of eventually dying from cancer or other disease caused by the pesticides and other residues and toxins" contained in the amount of produce they would typically eat in one year. The participants were requested to make separate estimates corresponding to eating only organically grown, or only conventionally grown, produce. These estimates were indicated by marking on a risk ladder (Fig. 23) which reports the estimated annual risk of death, averaged over the U.S. population or relevant subset, from a variety of diseases, accidents, and occupational and recreational activities. The use of the risk ladder forces the respondents to compare their estimates to estimates of other risks and should help them make more consistent numerical estimates of the risk.

Both conventional- and organic-food buyers estimate similar, small risks from eating only organically grown produce. Table 15 summarizes their responses. Both groups think the lifetime risk of dying from some disease caused by eating organically grown produce for one year is very small, on the order of about 1 in a million. This risk is comparable to that of dying from tuberculosis or viral hepatitis, in an airliner crash (during a single trip), or of accidental death from electrocution, tornado, flood, or lightning (Fig. 23). It is comparable to the estimated lifetime risk derived from the TD_{50}s, 1.4 x 10^{-5}, equivalent to an average annual risk of 0.2 in one million (assuming a 70-year lifetime).

Table 15

REPORTED ESTIMATED RISK FROM CONSUMING
ORGANICALLY GROWN PRODUCE

Risk from Organically Grown Produce (Per Million)	Organic-Food Buyers		Conventional-Food Buyers	
	Number	Cumulative Percentage	Number	Cumulative Percentage
0 to 0.9	12	60	15	65
1 to 9.9	0	60	4	83
10 to 49	5	85	3	96
50 to 99	2	95	0	96
At least 100	1	100	1	100
Lower quartile		0.1		0.1
Median		0.3		0.4
Upper quartile		16		7

In contrast, the groups differ dramatically in their reported estimates of the risk from eating conventionally grown produce, as summarized in Table 16. The conventional-produce buyers estimate the risk from eating conventionally grown produce as only marginally greater than the risk from eating organically grown produce, on the order of about one in a million. The organic-food buyers, however, think the risks of eating conventionally grown produce are much higher, on the order of 100 or 1,000 in a million. The median reported risk estimate, 850 per million, is somewhat greater than the annual risk of digestive-organ or respiratory cancer (about 500 per million each) and almost as great as the annual lung cancer risk for a one pack (or more) per day smoker (about 1,200 per million).

The estimated risk reported by organic-food buyers seems very high compared with other risks. The median estimate is higher than the estimated risk of digestive-organ cancer, which one might think would be the most likely type of cancer that might be caused by pesticide residues. One probably cannot prove that the risk is not this high, however. Organic-food buyers claim that pesticide residues and other factors may cause a broad range of adverse health effects in addition to cancer. Some of these effects may not yet be apparent, because widespread pesticide use has only developed within the last 40 years. Cumulative effects of lifetime pesticide residue consumption have not yet occurred.

Table 16

REPORTED ESTIMATED RISK FROM CONSUMING
CONVENTIONALLY GROWN PRODUCE

Risk from Conventionally Grown Produce (Per Million)	Organic-Food Buyers		Conventional-Food Buyers	
	Number	Cumulative Percentage	Number	Cumulative Percentage
0 to 0.9	1	5	14	61
1 to 9.9	2	15	0	61
10 to 49	2	25	9	100
50 to 99	1	30		
100 to 999	4	50		
1000 to 1499	3	65		
1500 to 1999	2	75		
2000 to 2499	4	95		
2500 to 3000	1	100		
Lower quartile	65		0.3	
Median	850		0.8	
Upper quartile	2200		14	

The difference between groups in the perceived risk from eating conventionally grown produce leads to a dramatic difference in the estimate of the annual risk avoided by consuming organically grown produce, reported in Table 17. The organic-food buyers' median estimate of the avoided risk is 800 in one million but the conventional-food buyers' median estimate is only one-half in one million. Of the six conventional-food buyers who think that conventionally grown produce is safer than organically grown, only one estimated a significant difference in risk, about 400 in one million. The other five all estimate the difference as no greater than 0.1 in one million.

Table 17

REPORTED ESTIMATED RISK AVOIDED BY CONSUMING
ORGANICALLY GROWN PRODUCE

Risk Avoided (Per Million)	Organic-Food Buyers		Conventional-Food Buyers	
	Number	Cumulative Percentage	Number	Cumulative Percentage
Less than 0			6	26
0 to 0.9	1	5	8	61
1 to 9.9	3	20	6	87
10 to 49	1	25	3	100
50 to 99	1	30		
100 to 999	5	55		
1000 to 1499	2	65		
1500 to 1999	2	75		
2000 to 2499	3	90		
2500 to 3000	2	100		
Lower quartile	58		-0.01	
Median	800		0.5	
Upper quartile	2100		5	

CHAPTER 16
WILLINGNESS TO PAY FOR RISK REDUCTIONS

Revealed-preference and contingent-valuation estimates of WTP to reduce food-borne risks are derived in this chapter. The revealed-preference estimate is based on the estimated premiums paid for organically grown produce and the estimated risk avoided by substituting organic for conventional produce derived above. The contingent-valuation estimate is based on reported WTP for organic produce and perceived risk as reflected in responses to the questionnaire.

Revealed-preference estimates are constructing using both types of risk estimates derived in the previous chapter: the risk indices, which rank foods according to their average weighted pesticide concentrations, and the estimated average lifetime risk of cancer from consuming small quantities of pesticide residues. The estimates based on the risk indices serve two purposes. First, they can be used to rank foods by the relative cost to the consumer of reducing pesticide risks by switching to organically grown produce. Second, they allow one to estimate bounds on consumer WTP to avoid pesticide residues based on observed choices between organically and conventionally grown items. The estimate based on the average cancer risk is an estimate of the implied value of life.

Naturally, all of the quantitative statements about willingness to pay for risk reductions that follow are conditional on the validity of the assumptions on which their calculation depends. As discussed earlier, each of the elements—the estimated premiums, the incidence of pesticide residues on foods, and the hazards of specific pesticides—is subject to great uncertainty. Both the premium and incidence estimates are subject to sampling variability. The premiums are estimated for the late spring/early summer season and further depend on the assumption that consumers are indifferent between the taste, appearance, size, and other characteristics of organic and conventional food items (that is, that the incremental taste function $\tau_i(.)$ is identically zero for all consumers and foods).[1] Organic produce is assumed to be free of pesticides and to not impose offsetting risks. If these assumptions do not hold, the estimates are biased and the qualitative statements based on them are inaccurate.

The estimated premium/risk index ratios p_i/r_i are presented in Table 18, together with the medians and quartiles of the distribution of premium/risk index ratios across foods. The foods are ranked in ascending order of the premium/Combined risk index ratio. Each premium/risk index ratio estimates the cost at which an individual can reduce health risks by substituting organic for conventional produce. For some foods (avocados, bananas, and onions) the estimated ratio is infinite because CDFA tests detected no pesticides on the items sampled. Hence consumers buy no reduction in pesticide intake by paying a premium for the organic variety. For others, the estimated premiums and premium/risk index ratios are negative. Some of these estimated premiums may be less than zero because of sampling error, although in one case (grapefruit) the t-statistic of -2.1 may allow rejection of this hypothesis.[2] If the premiums really are negative, everyone who buys these foods should buy organic items since one can reduce his risk and save money simultaneously.[3]

The existence of foods with negative premiums suggests that some of the maintained assumptions do not hold for these foods; for example, some people may find organic grapefruit less attractive or tasty. If organic grapefruit is as attractive to consumers as the conventional variety and is cheaper to produce, one would expect conventional growers to produce grapefruit without pesticides and fertilizer. Alternatively, the price of conventional grapefruit may be artificially high because of a U.S. Department of Agriculture marketing order to which organic grapefruit is not subject, or the estimate may be due to seasonal or random variation in prices.

Under the assumptions adopted, no one should buy either the conventional variety of a food for which the organic premium is negative or the organic variety of a food where the conventional type is less costly and contains no residues. Consequently, these anomalous foods are not considered further when describing the results for alternative risk indices below.

Table 18

ESTIMATED PREMIUM/RISK INDEX RATIOS

Food	Premium ($/lb.)	Risk Index(a)						
		Frequency	Concentration	Tolerance	Acute	Chronic	Chronic Sum	Combined
Avocado	-0.022	(b)	(b)	(b)	(b)	(b)	(b)	(b)
Cauliflower	-0.095	-1.426	-3.438	-3.202	-14.225	-13.543	-2.934	-13.875
Lemon	-0.084	-0.430	-1.560	-3.355	-0.495	-2.377	-1.248	-0.819
Grapefruit	-0.118	-0.143	-0.087	-0.610	-0.299	0.092	-0.044	-0.141
Leaf lettuce	0.014	0.012	0.010	0.054	0.003	0.014	0.004	0.005
Potato	0.026	0.043	0.013	0.203	0.333	0.011	0.004	0.021
Cucumber	0.149	0.155	0.069	0.004	0.022	0.040	0.017	0.028
Valencia orange	0.068	0.050	0.023	0.251	0.151	0.018	0.008	0.032
Romaine lettuce	0.022	0.065	0.083	0.325	0.042	0.178	0.040	0.068
Spinach	0.221	0.191	0.208	0.073	0.057	0.841	0.170	0.110
Apricot	0.148	0.845	0.422	4.224	2.851	0.241	0.106	0.445
Kiwi	0.164	0.435	0.228	3.404	0.687	0.330	0.130	0.446
Red cabbage	0.218	0.752	0.909	3.645	0.288	2.110	0.562	0.507
Green cabbage	0.220	0.760	0.919	3.684	0.291	2.133	0.568	0.513
Green pepper	0.434	0.845	1.336	2.387	0.472	1.315	0.578	0.694
Celery	0.286	0.501	0.420	4.038	0.770	1.001	0.263	0.870
Peach	0.674	0.489	0.340	1.809	2.368	0.686	0.159	1.064
Broccoli	0.625	2.152	2.455	11.665	3.954	1.695	0.810	2.372
Carrot	0.245	1.407	3.862	2.993	2.006	4.253	1.230	2.726
Bing cherry	0.948	4.545	2.273	45.455	91.827	9.090	1.136	16.543
Apple	0.371	8.258	7.682	55.066	9.447	125.786	55.066	17.575
Zucchini	0.372	2.353	33.659	14.134	25.387	29.044	9.365	27.093
Tomato	1.162	4.244	19.205	26.738	37.383	51.921	13.864	43.478
Yellow squash	0.374	5.164	33.289	16.647	32.884	101.833	35.638	49.727
Red onion	0.039	(b)	(b)	(b)	(b)	(b)	(b)	(b)
Banana	0.117	(b)	(b)	(b)	(b)	(b)	(b)	(b)
Spanish onion	0.385	(b)	(b)	(b)	(b)	(b)	(b)	(b)
Lower Quartile(c)	0.122	0.155	0.083	0.251	0.151	0.178	0.040	0.068
Median(c)	0.221	0.752	0.422	3.404	0.687	1.001	0.263	0.220
Upper Quartile(c)	0.383	2.152	2.455	11.665	3.954	4.253	1.136	2.726

(a) Ratios in dollars per pound per unit of risk index. Risk-index units are defined in Table 14.
(b) Commercially grown items are estimated to contain no residues. Hence estimated willingness to pay is infinite (minus infinity for avocadoes, since estimated premium is less than zero).
(c) The order statistics are taken over the foods with finite positive premium/risk index ratios.

RELATIVE COSTS OF AVOIDING PESTICIDE RISKS ACROSS FOODS

One use of the estimated premium/risk index ratios is to distinguish foods where the cost of risk reduction by switching from conventional to organic produce is small from those where it is large. The ranking of the foods by premium/risk index ratio is not very sensitive to the specific risk index used since all of the risk indices are highly correlated. As shown in Table 19, most of the Spearman rank correlation coefficients between risk indices are greater than 0.8 or 0.9. The smallest correlations are between the Acute risk index and the two measures of chronic risk, but even these exceed 0.7. Thus the choice of how to combine currently available toxicological data with information on pesticide incidence is apparently not critical, although one's willingness to trade between acute and chronic risk could have some effect on his choice of foods.

The foods with relatively extreme premium/risk index ratios are quite consistent across risk indices. Four foods are among the six with estimated premium/risk index ratios less than or equal to the lower quartile using any of the seven risk indices. These foods, for which the estimated cost of risk reduction is fairly small, are red and green leaf lettuce, cucumbers, and valencia oranges. Potatoes and romaine lettuce have estimated ratios in the lowest quartile under six of the risk indices and spinach ranks in the lowest quartile using two of the indices. Similarly, the foods with the highest estimated cost of reducing pesticide risks are consistent across risk indices. Again, four of the foods rank in the top six using

Table 19

SPEARMAN RANK CORRELATION COEFFICIENTS
AMONG RISK INDICES

Risk Index	Fre-quency	Concen-tration	Toler-ance	Acute	Chronic	Chronic Sum	Combin-ation
Frequency	1.0	.940	.944	.827	.887	.920	.935
Concentration		1.0	.887	.733	.957	.989	.930
Tolerance			1.0	.858	.848	.876	.926
Acute				1.0	.707	.712	.887
Chronic					1.0	.973	.915
Chronic Sum						1.0	.927
Combination							1.0

each of the indices: yellow squash, tomatoes, zucchini, and apples. Celery ranks in the highest quartile under six of the indices and broccoli and carrots rank among the highest under four indices. Thus, if one wishes to reduce his or her risk from ingesting pesticides without switching completely to organic produce, he or she should switch to organically grown versions of the foods that have low estimated premium/risk index ratios, like lettuce and spinach, and stay with the conventional versions of foods with relatively high ratios, like squashes, celery, and carrots.

REVEALED WILLINGNESS TO PAY TO REDUCE PESTICIDE RISKS

The estimated premium/risk index ratios can also be used to estimate bounds on a consumer's WTP to avoid pesticide residues based on his or her choice between organic and conventional produce. As discussed in Chapter 13, the theoretical consumer will purchase the organic type of each food for which the premium/risk ratio, or price of risk reduction, is smaller than his WTP (that is, foods for which the premium/risk ratio is less than the critical value λ^*) and the conventional type of all other foods. An individual who buys only conventional produce implicitly reveals that he or she is unwilling to pay at a rate equal to the lowest estimated premium/risk index ratio (of the foods consumed) to reduce pesticide intake. Similarly, one who buys only organic produce implicitly reveals a WTP at least as great as the highest of the ratios estimated for the foods bought.

Since many consumers buy either all conventionally or nearly all organically grown produce, one might infer that their WTP is outside the range of ratios I estimate. However, this inference is too strong. As confirmed by the focus groups, consumers do not have enough information about the relative risks of foods to choose the optimal mix of organic and conventional items. Most believe the risk avoided by consuming organically grown produce is about the same for all types of produce. Consumers may further simplify their decisions by acting as if they believe that the premium/risk ratios are about the same for all foods, and thus choose either all organic or all conventional items. If an individual's WTP is such that only a few organically grown items are worth purchasing, it may not be worthwhile to incur the costs of travelling to one of the stores that offer organic produce if a conventional market is more convenient. Finally, the estimates of the extreme premium/risk index values are biased towards the extremes because of sampling variability.[4]

A crude adjustment for these problems is to use the quartiles of the estimated distribution of premium/risk index ratios (reported in Table 18) to estimate bounds for WTP. For example, for the premium/Concentration risk index ratio the quartiles are 0.083 and 2.455 dollars per pound per part per million of residues avoided. These statistics suggest that individuals who buy only organic produce implicitly reveal that their WTP to avoid pesticide residues is at least $2.45 per pound per part per million of residues. Individuals who buy only conventionally grown produce implicitly reveal their WTP is less than 8 cents per pound per part per million. The other premium/risk index ratio estimates can be interpreted similarly.

REVEALED-PREFERENCE ESTIMATES OF THE VALUE OF LIFE

The lifetime risk of cancer from ingesting the pesticide residues found on conventionally grown produce was crudely estimated as about 10 in one million (Chapter 15). If it is assumed that the quantities of pesticides found on organically grown produce are negligible in comparison and that organically grown produce does not impose offsetting risks, an approximate value of life implicit in consumers' willingness or unwillingness to pay the market premiums for organically grown produce can be estimated. This estimate is simply the ratio of the lifetime additional cost of buying organic instead of conventional produce to the incremental risk reduction. As an estimate of the risk avoided by switching to organically grown produce, use 1.4×10^{-5}. This is the estimate based on the median TD_{50} of the pesticides for which TD_{50}s are reported. Although it may be too high an estimate for the average quantity-weighted pesticide, its use may roughly account for the other potential health consequences of pesticide residues. For comparison, the implied value of life using the high estimate of avoided risk, 1.1×10^{-3} (based on the Crouch and Wilson (1979) estimate of the carcinogenic potency of EDB) is also calculated.

Because the additional expenditure is distributed over the consumer's lifetime, it is not appropriate to simply sum the annual premiums. Instead, I use the present value of the lifetime stream of additional payments, assuming the individual chooses (once and for all) whether or not to purchase organically grown produce at age 20 and lives to age 70.[5] For individual decision making, the appropriate discount rate is based on the individual's incremental cost of funds to be spent on organic produce. The incremental funds will come from money that the consumer would otherwise save or spend on other goods. For most consumers the opportunity cost of these funds is equal to the rate of return that could be earned on savings, net of inflation and taxes. It may be greater if the consumer's budget constraint is so tight that he or she has no savings and the next best marginal use of funds (other than incremental spending for organic produce) is more highly valued than the best investment opportunity. The incremental cost of buying organic produce is consequently the present value of the future stream of additional payments discounted at the rate at which the consumer could earn income from his savings (or higher for consumers without savings).[6]

The annual premium is estimated using the quartiles of the premium/Chronic risk index ratio. For example, the median of this ratio is $1 per pound per part per million of carcinogens avoided. An individual who purchases almost all organic produce thereby reveals his willingness to pay at this rate (or higher) to reduce his ingestion of carcinogens. The average concentration of potentially carcinogenic residues, as measured by the Chronic risk index, is 0.1 ppm. Assuming, as in the risk calculations, that a typical individual consumes 1 kg (2.2 lb.) of produce daily, the annual premium such an individual would pay to consume organic instead of conventional produce is about $80 ($1/lb.-ppm residues avoided x 2.2 lb. x 0.1 ppm avoided x 365 days/year). Assuming a 5-percent discount rate, the present value of a 50-year stream of annual payments of $80 is about $1,400. Dividing by the estimated avoided risk of 1.4×10^{-5}, the individual's implicit value of life is about $100 million (= $1,400/$1.4 \times 10^{-5}$).[7]

Table 20 displays a range of calculated implicit values of life assuming alternative discount rates, levels of WTP for organically grown produce, and avoided risk. The estimated implicit values of life can be interpreted as follows: An individual who buys no organically grown produce implicitly reveals that he is unwilling to pay at a rate equal to the lower quartile of the distribution of the premium/Chronic risk index ratio to avoid the possibly carcinogenic pesticide residues contained in conventional produce. Assuming that the avoided risk is 1.4×10^{-5} and that the individual can earn a real return of 5 percent on his savings, net of tax, he implicitly reveals that he values his life at no more than $19 million. Similarly, an individual who buys practically all organically grown produce reveals that he is willing to spend at a rate at least as great as the highest quartile of the premium/Chronic risk index ratio, and thereby reveals that he implicitly values his life at more than $440 million.

Compared with other estimates of the implicit value of life reported in the literature, the estimates based on the "best guess" of the risk avoided by substituting organic for conventional produce are high. Fisher et al. (1986) review the major studies of implied valuation of life and conclude that the best estimate is a range of $1.6 to $8.4 million. The

Table 20

ESTIMATED IMPLICIT VALUES OF LIFE UNDER ALTERNATIVE ASSUMPTIONS

(Millions of dollars)

Discount Rate (percent)	Willingness to Pay to Reduce Risk		
	Lower Quartile (-0.18)	Median (-1.00)	Upper Quartile (-4.25)
Avoided Risk = 1.4×10^{-5} (Best Guess)			
0.0	52	290	1,200
2.0	32	180	780
5.0	19	100	440
10.0	10	57	240
Avoided Risk = 1.1×10^{-3} (High Estimate)			
0.0	0.66	3.7	16
2.0	0.41	2.3	9.7
5.0	0.24	1.3	5.7
10.0	0.13	0.72	3.1

Note: The three values of willingness to pay for risk reductions are the quartiles of the distribution of the premium/Chronic risk index ratio, measured in dollars per pound per part per million of pesticide residues weighted by the Chronic risk index weighting factors.

fact that most consumers do not find it worthwhile to purchase organically grown produce is not surprising, given the apparently very high value of life implicit in such a decision. In contrast, the estimates using what I consider a very high estimate of the avoided risk, an estimate that assumes that all of the potentially carcinogenic pesticides (as measured by the Chronic risk index) are as potent as EDB, produces value of life estimates quite consistent with studies of risk taking in other areas. Most of the organic-food consumers in the focus groups claim that the avoided risk is even higher than this estimate, and consequently that the lower bound on the value of life implicit in their choice of organic produce is even smaller than shown in the bottom panel of Table 20.

CONTINGENT-VALUATION ESTIMATES OF THE VALUE OF LIFE

The contingent-valuation estimates of organic-food buyers' WTP for organic produce coincide closely with the premiums estimated from market prices. Hence, the contingent-valuation estimate of the value of life based on reported WTP and estimates of the avoided risk are comparable to the revealed-preference estimates. An alternative contingent-valuation estimate can be derived from reported WTP for organic produce and the consumers' own estimates of the avoided risk, reported through the survey. The distribution of these values is presented in Table 21.[8]

The median values are comparable to the range estimated in other areas, $1.6 to $8.4 million (Fisher et al., 1986). The values for the organic-food buyers are an order of magnitude smaller than for the conventional-food buyers. One might expect, at least with regard to food purchases, the organic-food buyers to act as though they implicitly value their lives higher than conventional-food buyers. Although the organic-food buyers are willing to pay much more for organically grown produce, they estimate the risk of the conventional produce as so much higher than do the conventional-food buyers that their implied values of life are smaller. A number of potential explanations for this outcome are possible.

First, reported WTP may be biased. The organic-food buyers' estimates may be biased downward because of "anchoring" to the actual premiums they observe in the market. In fact, they might be willing to pay more for organically grown produce if necessary. Several suggested that the possibility of growing one's own produce organically served as an upper bound on the premium they would pay. Similarly, the conventional-food buyers' reported WTP may be too large. Most of these consumers think the difference in risk between the types of produce is almost negligibly small. They believe they should be willing to pay some additional premium for the safer type but may have difficulty distinguishing between fairly small premiums, for example between 1 and 5 percent. Their assessments might have been improved if they had been asked for WTP in dollars per week or some other absolute rather than relative metric. However, any bias in the median resulting from this "round-off error" might be roughly counterbalanced by the five respondents who reported an incremental WTP of zero, presumably because they think the difference in risk is too small to worry about.[9]

Second, the reported risk reduction from avoiding conventionally grown produce may be over-stated. One wonders whether the organic-food buyers would stand by their estimates if they were pushed to defend them in comparison with other risks. Their

Table 21

VALUE OF LIFE IMPLIED BY REPORTED WILLINGNESS TO PAY AND ESTIMATED AVOIDED RISK

Implied Value of Life ($ million)	Organic-Food Buyers		Conventional-Food Buyers	
	Number	Cumulative Percentage	Number	Cumulative Percentage
0	0	0	5	23
0 to 0.1	2	10	1	27
0.1 to 1	9	55	0	27
1 to 10	3	70	6	55
10 to 100	4	90	5	77
100 to 1,000	1	95	4	95
Over 1,000	1	100	1	100
Lower quartile	0.25		0.05	
Median	0.75		6.62	
Upper quartile	21.6		93.5	

Note: A sixth conventional-food buyer reported a willingness to pay of zero but also an estimated risk difference of zero. The calculated implicit value of life is indeterminate.

responses could have been biased upwards by the initial discussion of food risks, but this seems unlikely because the conventional-food buyers' responses do not seem to be biased upwards to the same extent, if at all.

Finally, the calculated implicit value of life for the conventional-food buyers is highly sensitive to any round-off error or other inaccuracy in the reported risk difference, because the estimated difference is so small. Estimates might have been lower had even smaller risks been included on the risk ladder. The median reported difference is only 0.5 in one million; if it were one in a million the median implicit value of life for this group would fall to $3 million.

CONSISTENCY WITH MANAGEMENT OF OTHER HEALTH RISKS

Economic theory suggests that individuals should manage the array of risks they face consistently: They should allocate resources to reducing each risk until the implicit price of further marginal reductions exceeds the value of life. Research on seat-belt use has shown that use is correlated with management of other personal risks: Seat-belt use is lower among heavy drinkers, smokers, overweight and physically inactive motorists (Goldbaum et al., 1986; Mayas et al., 1983). To explore the consistency with which consumers manage risks, focus-group participants were asked whether they smoke cigarettes, wear automobile seat belts, eat meat, and drink bottled water.

Except for seat-belt use, the organic-food purchasers appear to be more health conscious than the conventional-food purchasers. The organic-food buyers are less likely to smoke cigarettes (5 vs. 13 percent), more likely to have reduced their consumption of red meat (100 vs. 55 percent), and much more likely to drink bottled instead of tap water (70 vs. 18 percent always drink bottled water, 100 vs. 27 percent drink bottled water usually or always). Although none of the contrasts are statistically significant, in part because of the small sample, they are suggestive.

Organic-food purchasers appear less likely to wear seat belts, however.[10] As summarized in Table 22, for any specified frequency a larger fraction of conventional- than organic-food buyers report they wear their seat belts at least that often.

Again, the difference is not statistically significant. If organic-food purchasers are in fact less likely to wear seat belts, there are a number of possible although not necessarily convincing explanations involving the differences between the risk of death or injury through an automobile accident and the risk of disease through consumption of various substances. Several of the organic-food purchasers recognized this difference, claiming that the risks are "not the same thing." For example, the possibility of dying a slow death from a degenerative disease may be more frightening than a relatively quick death due to an automobile accident (although automobile accidents more often maim than kill). Probably a more important factor is that people are typically less concerned about risks over which they believe they have some degree of control, such as motor-vehicle accidents (Kahneman et al., 1982). A great majority of drivers—75 to 90 percent in one study reported by Slovic et al. (1978)—believe themselves to be better than average drivers. The risk of an automobile accident during a single trip is resolved quickly, so anxiety is short-lived. One knows at the end of the trip whether he survived or not. In contrast, because food-borne hazards may take

Table 22

REPORTED FREQUENCY OF SEAT-BELT USE

Frequency of Use	Organic-Food Buyers		Conventional-Food Buyers	
	Number	Cumulative Percentage	Number	Cumulative Percentage
Always	6	27	11	48
Usually	8	64	6	74
Sometimes	2	73	2	83
Rarely	3	86	2	91
Never	3	100	2	100
Mean	2.5		2.0	

Note: Mean calculated using a five-point scale: Always = 1, Usually = 2, Sometimes = 3, Rarely = 4, Never = 5.

many years to manifest one may experience continual anxiety. Finally, individuals may simply be concerned with different risks: some are fearful of ingesting harmful substances, others of automobile or other accidents.[11]

Risk management decisions are at least sometimes correlated with factors that should increase one's implicit value of life. Seat-belt use is higher among wealthier and better educated motorists; it is positively correlated with age (Goldbaum et al., 1986; Haaga, 1986; Leigh, 1990; Lund, 1986; O'Neill et al., 1983). I analyzed the correlations between reported management of other risks and calculated the implicit value of life within groups of organic- or conventional-food buyers. Few correlations are evident, although the correlations that do appear are consistent with expectations. For example, among the organic-food purchasers, the median implicit value of life is higher for those who always or usually use seat belts than for the others. Similarly, conventional-food buyers who have reduced their red meat consumption have higher average implied values of life than those who have not.

In summary, the implicit values of life calculated from reported WTP for organically grown produce and beliefs about the risks from consuming organically and conventionally grown produce are approximately consistent with those calculated in other contexts. Surprisingly, the organic-food buyers' implied value of life is smaller than the conventional-food buyers'. Although the organic-food buyers report a much higher WTP, their estimates of the risks from consuming conventionally grown produce are so great that their implicit values of life are smaller than those of the conventional-food buyers. Several potential explanations for this result, involving biased or inaccurate reporting of WTP and risk avoided, have been suggested. As one might expect, the organic-food buyers appear to be more health conscious in their management of other risks involving ingestion of various substances but they may be less cautious in managing the risk of motor vehicle accidents. While these results are suggestive, they are derived from a limited exploratory study based on a small and unrepresentative sample.

CHAPTER 17
CONCLUSIONS

Because many of the benefits of environmental policies occur in forms that are not traded on economic markets, it is not possible to value them by referring to their market prices. Two classes of methodologies—revealed preference and contingent valuation—have been developed to estimate benefits. Each has advantages and disadvantages.

Revealed-preference methods have the advantage of incorporating actual behavior, so they are not subject to the criticism of being based on responses to hypothetical questions. However, they cannot be developed unless preferences for the environmental attribute in question can be inferred from observable behavior in a related market, a potentially severe constraint. In contrast, contingent-valuation estimates can always be obtained by carefully specifying the condition to be valued and surveying individuals to obtain their self-reported values. Their weakness is that individuals may not respond accurately. Determining one's WTP for an environmental attribute or health-risk reduction can be a difficult cognitive task, in part because it may involve evaluating choices one rarely faces explicitly; one may not even know his preferences for situations far removed from his usual context. Individuals may find assigning a monetary value to environmental attributes to be offensive; alternatively, they may attempt to misrepresent their preferences for strategic reasons, although experience with the contingent value method suggests that misrepresentation is not a serious concern.

In this section, I have estimated consumer WTP to reduce food-borne risks using each method. The revealed-preference and contingent-valuation estimates of WTP for organically grown produce are comparable. This is perhaps unsurprising, since reported WTP may be tied to knowledge of the premium organic-food buyers do pay. A possible weakness in these estimates is that WTP for organic produce may include other perceived benefits in addition to the supposed health improvement, although the focus-group discussants suggested that health risks were at least the most important difference between produce types.

Estimates of WTP to reduce risk, frequently characterized as a "value of life," are difficult to obtain in this context because of substantial uncertainty about the relationship between pesticide consumption and health outcomes, and about the exact differences in pesticide contamination and other factors between produce types. To accommodate this outcome uncertainty I applied a variety of risk estimates ranging from "risk indices" that combine pesticide incidence and toxicity information in diverse forms, to estimates of the lifetime cancer risk using dose-response information for selected compounds, to consumer estimates of the difference in risk. The value of life estimates using the incidence and toxicity data are larger than reported in other risk contexts, supporting the hypothesis that organic-food purchasers are an extreme group and that most consumers' failure to buy organic produce does not imply they are failing to efficiently allocate resources to reducing health risks. Based on their reported estimates of the health risks, however, the reason organic-food buyers make that choice does not appear to be because they assign a higher value to risk reduction; they apparently believe the risk avoided by consuming organic produce is much larger than other consumers estimate it to be. In contrast, conventional-food buyers report that they do not give much attention to the possible adverse health effects of pesticides: They rely on government agencies and supermarkets to protect them.[1]

The disagreement among consumers about differences in health risk suggests that improved public information concerning food-borne risks might lead to more accurate risk perceptions and improve consumers' ability to efficiently minimize the health risks they face. Unfortunately, the toxicological and epidemiological data and biological understanding necessary to accurately characterize the health risks of cumulative subacute ingestion of pesticides are not currently available. In the interim, the risk indices developed here may be useful as easily applied, proxy measures of risk. They could be used to help inform consumers about pesticide incidence and toxicity, to allow them to more efficiently reduce their risk from ingesting pesticide residues. Moreover, the risk indices may be used to help allocate testing and epidemiological resources to the foods and pesticides for which improved information would be most valuable and to help decide which legal tolerances should be reassessed.

PART III

EXTENSIONS:
GLOBAL CLIMATE CHANGE

CHAPTER 18

APPLYING THE TOOLS TO POTENTIAL CLIMATE CHANGE

Several techniques for dealing with the problems of uncertainty about outcomes and values have been described and applied in this book. Both to summarize and to extend this work, I discuss how these tools can be applied to another environmental policy issue of current interest, the possibility of global climate change.

GLOBAL CLIMATE CHANGE

Global climate change is often called "global warming" or the "greenhouse effect." Essentially, anthropogenic increases in the atmospheric concentrations of several trace gases are likely to produce shifts in the earth's climate, including an average warming. Current estimates suggest that the global average air temperature near the earth's surface is likely to increase between about two and five degrees centigrade sometime in the next century if concentrations of the greenhouse gases continue to climb. A higher average temperature is likely to be accompanied by other climatic changes, including changes in the quantity and geographical distribution of precipitation and the frequency and intensity of storms and droughts. Hotter and drier conditions could lead to more frequent and severe fires. The sea level would likely rise (perhaps a meter or so over the next century) as the oceans warm and expand, and could rise much more if large glaciers on land (principally in Greenland and Antarctica) melt into the seas.[1]

The Greenhouse Effect

It is well established the the earth's atmosphere currently produces a "greenhouse" effect by inhibiting the radiation of heat from the earth's surface to space.[2] In the absence of the atmosphere, the earth's surface temperature is calculated to be -18 degrees centigrade, about 33° C lower than the current average of about 15° C (59 degrees Fahrenheit). The greenhouse effect occurs because most of the incoming solar energy is contained in the ultraviolet, visible, and near-infrared portions of the electromagnetic spectrum, to which the atmosphere is relatively transparent. The earth's surface radiates heat back to space, but much of this energy is contained in the infrared portion of the spectrum. Water vapor and other atmospheric constituents absorb much of the energy in this region, leaving only a few narrow "windows" that account for much of the escaping infrared radiation (Ramanathan, 1988). The "greenhouse gases" or GHGs are trace gases that absorb strongly in these radiation bands. By further blocking outgoing infrared radiation and re-radiating it in all directions, GHGs limit the escape of heat energy from the earth-atmosphere system. Increases in GHG concentrations may further reduce the energy radiated to space, and so warm the planet. Because the energy radiated by infrared emissions increases with the earth's temperature, equilibrium is re-established when the temperature rises enough that the energy radiated to space, net of atmospheric absorption, equals the incoming solar energy.

It is useful to divide the greenhouse effect between forcing and feedback effects. The greenhouse forcing can be measured as the increased surface temperature necessary to re-establish an equilibrium between incoming solar and outgoing radiant energy. The equilibrium temperature increase associated with a prescribed increase in greenhouse forcing is believed to be several times larger than that associated with this black-body radiation effect, because of positive feedbacks in the climate system.

The most important feedback mechanisms are believed to include increases in water vapor, decreases in the surface albedo, and changes in cloud cover and radiative properties. The increase in water vapor (evaporated from the oceans) is likely to occur because warm air can hold more water vapor than cold. Because water vapor is also a GHG, this would enhance the direct forcing. The change in albedo would result from melting sea ice and decreases in the land area covered by snow. While snow and ice reflect a relatively large fraction of solar radiation, the darker sea and ground that would be uncovered would absorb more of this heat energy. The feedback effect of clouds is less certain. Clouds reflect incoming radiation, absorb outgoing radiation, and re-emit it at a temperature that varies with the height of the cloud. Current understanding suggests that increased greenhouse forcing would produce less cloud cover over low and middle latitudes (where solar insolation is greatest) and higher cloud tops (which emit less energy because of their lower temperature), but these positive feedback effects might be offset by changes in cloud composition (Mitchell, 1989; Raval and Ramanathan, 1989; Ramanathan et al., 1989).

The major GHGs of concern include carbon dioxide (CO_2), methane (CH_4), nitrous oxide (N_2O), ozone (O_3), and the potential ozone depleters, principally CFCs 11 and 12. Of these, CO_2 accounts for about half the current increase in greenhouse forcing relative to pre-industrial conditions. The atmospheric concentrations, sources, and sinks of CFCs 11 and 12 and the other PODs are described in Part I; the other GHGs are discussed below.

The concentration of CO_2 is currently about 350 parts per million (ppm), having increased from about 315 ppm in 1958 (when reliable measurements began) and from an estimated value of about 275 ppm in pre-industrial times (Mitchell, 1989). At present, the CO_2 concentration is increasing about 1.5 ppm (0.5 percent) per year.

Atmospheric CO_2 is part of the global carbon cycle. It is released to the atmosphere through respiration by plants and animals, and is removed and transformed into cellulose and other materials by plants through photosynthesis. CO_2 is also absorbed by the oceans and is incorporated into plankton that die and eventually settle to the bottom, incorporating the carbon in carbonaceous compounds (Brewer, 1983). Only about half of the carbon calculated to have been released by fossil-fuel combustion since 1958 remains in the atmosphere. Oceanic uptake is believed to account for the other half, although quantitative understanding of these mechanisms is not clear (Mitchell, 1989), and the possibility of significant terrestrial sinks has been suggested (Tans et al., 1990).

The primary sources of the increased atmospheric CO_2 are believed to be anthropogenic: combustion of fossil fuels and the loss of carbon from forests that are cleared for agricultural or other uses. Since fossil fuels represent large stores of carbon removed from the atmosphere by plants and buried, their combustion is releasing the incorporated carbon at a much quicker rate than it is absorbed by new plant growth. The net deforestation that is occurring, particularly in tropical areas, reinforces the effect of fossil-fuel combustion by reducing the net CO_2 uptake by the biosphere. In addition, as forests are cleared the plant materials are often burned or left to decay, releasing much of the stored carbon to the atmosphere (Detwiler and Hall, 1988).

Methane concentrations have been increasing at about 1 percent annually, and currently equal about 1.7 ppm. The major sink is believed to be tropospheric oxidation by the hydroxyl radical (OH); sources include fossil fuel combustion and leakage (natural gas is primarily methane), emissions by anaerobic bacteria living in the gut of ruminants (e.g., cattle, sheep), insects (e.g., termites), rice paddies, and natural wetlands (Blake and Rowland, 1988; Pearman and Fraser, 1988; Lowe et al., 1988). Warming in polar regions could create a positive feedback if it produced larger CH_4 fluxes from tundra, although recent evidence suggests a negative feedback associated with dryer soils (Whalen and Reeburgh, 1990).

Understanding of observed increases in nitrous oxide concentrations is also limited. The concentration of N_2O is about 300 parts per billion (ppb) and is currently increasing at about 0.2 percent annually. Its primary sources are believed to include nitrogen-containing fertilizers, microbial processes in soil and water, and fossil-fuel combustion. The principal sink is apparently photolysis with atomic oxygen in the stratosphere.

The role of ozone is complicated: in the troposphere, ozone acts as a greenhouse gas, but in the stratosphere it produces a net cooling by absorbing incoming ultraviolet radiation. About 90 percent of atmospheric ozone is concentrated in the stratosphere, where there is some evidence of an average decrease (Watson et al., 1988). In the troposphere, ozone is an important constituent of photochemical smog. It is formed by reactions involving NO_2 and hydrocarbons in the presence of sunlight. Average concentrations appear to have increased during the last century (Hough and Derwent, 1990). Because its residence time is measured in hours or days, the concentration of tropospheric ozone varies substantially by location and time (Seinfeld, 1989; Penner, 1990)

Mathematical Climate Models

The factors influencing global climate change are varied and complex. They include changes in the atmospheric concentration of a number of greenhouse gases, volcanic aerosols, and other constituents, changes in the absorption and transmission of electromagnetic radiation by the atmosphere, changes in convection, cloud formation, snow and sea-ice cover, horizontal and vertical heat transport by winds and ocean currents, and possible biological feedbacks. In order to integrate as many of the presumably important factors as possible, scientists rely on computer-based models of the climate system.[3]

Early models were either energy-balance or radiative-convective models. Energy-balance models calculate the characteristic temperature of the planet as a function of solar insolation, planetary albedo, and infrared absorption. They may treat the earth as a point (zero dimensional) or represent the change in temperature with latitude (one-dimensional). Radiative-convective models refine the energy-balance models by explicitly describing the vertical temperature profile of the atmosphere, incorporating heat transfer by infrared radiation between atmospheric layers and the surface and turbulent heat-transfer processes involving both dry and moist convection.

General Circulation Models (GCMs) define the current state of the art in climate modeling. GCMs have been developed from models used to forecast weather over periods of several days. The models are three-dimensional, and calculate the temperature, wind velocity, water-vapor concentration, cloud cover, surface pressure, snow mass, soil temperature and moisture content on a grid of points representing the earth's surface. GCMs include a set of nonlinear partial differential equations describing the heat and motion fields of the atmosphere. These equations are solved (approximately) on a finite set of grid points at discrete time intervals. Current models divide the atmosphere into about ten vertical layers and use horizontal grid spacings of a few hundred kilometers; the time step is typically between 10 and 40 minutes. To the extent possible given the model resolution, the geography and orography are realistic. Because of their complexity, the models are run on the fastest available computers; even so, GCMs typically require between one half and a few minutes to simulate the global weather conditions for a single day (Schlesinger, 1984).

Important weaknesses in current GCMs include the representation of clouds and of the oceans. Clouds and other phenomena that occur on subgrid scales are parameterized, or represented by coefficients that characterize the extent to which that phenomenon is calculated to occur within the region represented by a grid point (e.g., the fraction of the sky covered by clouds). The detailed physics of cloud formation and dissolution cannot be modeled within current GCMs.

The treatment of heat uptake and transport by the oceans is not important for simulating periods of several days, for which the GCMs were originally developed. For changes occurring over decades or longer, however, the thermal inertia of the oceans and redistribution of heat is believed to be important. Early GCMs incorporated a "swamp ocean," one that was always in thermal equilibrium with the local atmosphere and could not transport heat horizontally. The disadvantage of such an approach is that the simulated oceans would freeze at night and in the winter in polar regions, so simulations using a swamp ocean cannot include diurnal and seasonal variations in local solar insolation. More recently, models have incorporated "slab model" oceans. In this case, only a thermally mixed upper layer of the oceans is included, typically on the order of 100 m deep. This upper layer may transmit heat diffusively to the deep oceans, but up- and downwelling and climatically-induced changes in oceanic heat transport are not included (Hansen et al., 1988). General circulation models of the oceans, which describe oceanic circulations using partial differential equations based on physical principles, also exist and have been linked to atmospheric GCMs, although the linked models are not yet as refined (Mikolajewicz et al., 1990; Schlesinger, 1984).

In attempting to characterize the changes in climate that may result from increasing GHG concentrations, GCMs may be used in either equilibrium or dynamic studies. In equilibrium studies, the model is run for a sufficiently long period that the simulated climate (frequency distribution of weather) stabilizes. Typically, equilibrium results are obtained for the current or pre-industrial atmospheric composition (the control run) and for a benchmark altered atmosphere, typically one with twice the pre-industrial CO_2 concentration (although higher CO_2 levels have also been simulated). Comparison of the two runs provides information about how the future climate, once stabilized, might compare with the present.

In a dynamic study, the concentration of GHGs is altered over time in a prescribed manner corresponding to some plausible emissions scenario. Again, the GCM simulates the weather at approximately half-hour intervals around the globe; climate summaries are based on comparing the frequency distribution of simulated weather between different periods of the simulation. Because the thermal inertia of the oceans, the cryosphere (ice and snow), and other components of the climate system may have important influences on the rate of climatic change, the realism of dynamic results is necessarily tied to the adequacy with which these non-atmospheric processes are modeled.

GCMs attempt to predict climate by predicting geographically and temporally detailed weather. It has been recognized for some time that detailed weather prediction over periods of more than a few months or so may be impossible, because of inaccuracies in observations of the initial conditions with which the simulation begins, as well as in the equations and solution algorithms used. The general problem of the sensitive dependence of a simulation result on errors in measurement of initial conditions is studied under the rubric of "chaos theory" (Lorenz, 1969a, 1969b; Gleick, 1988; Stewart, 1989). The use of GCMs to simulate future climate assumes that, even though the detailed weather simulations are not reliable, the climate defined by the simulated weather is predictable. Although climate modelers claim that the correspondence between simulation results and observations of climatic variations on seasonal and geological time scales provide evidence of model reliability, others argue that there does not exist any adequate theory concerning the predictability of climate (Somerville, 1987).

Consequences of Climate Change

The fundamental characteristic of global climate change is expected to be an average warming. However, the change in temperature is not likely to be uniform in space or season; at present, it is thought that the greatest warming will occur at high latitudes and during the winter. The average and differential warmings are likely to change regional and global atmospheric and oceanic circulation patterns, further altering local seasonal patterns of temperature and precipitation. The details of such changes are not clear, however, with different models predicting differing effects. Most state-of-the-art models seem to predict increased precipitation at high latitudes, especially in the winter; but the models disagree on whether the Southeast Asian monsoon will become stronger or weaker (Mitchell, 1989). A possible consequence of differential warming would be the weakening of oceanic circulations that distribute heat from tropical to temperate latitudes, with potentially dramatic effects on regional climate. For example, it has been suggested that the current pattern of circulation in the North Atlantic (including the Gulf Stream) is not the only stable pattern, and that in other geologic eras this circulation was much weaker or possibly opposite in direction. If global warming were to upset existing circulation patterns, the climate in Europe could be much colder than at present (Broecker et al., 1985; Broecker, 1987).

Other potential consequences include changes in the seasonal and geographic distribution of precipitation. Combined with more rapid snowmelt and enhanced evapotranspiration, these changes could disrupt current water supply arrangements and reduce the production of hydroelectric power through existing systems. In particular, it has been suggested that mid-continent areas, such as the U.S. grain belt, might become much more dependent on irrigation, and that spring runoff would be inadequate for satisfying Western U.S. requirements (Revelle and Waggoner, 1983; Gleick, 1989).

A major result of global warming could be a rise in sea level due to thermal expansion as the oceans warm and to net melting of land ice, especially in Greenland and Antarctica. In addition to coastal flooding (particularly under storm-surge conditions), sea-level rise could promote saline infiltration of fresh-water aquifers.

Although natural ecosystems could potentially migrate to areas of suitable climate and/or adapt to climate change in their current locations, the pace of climate change may far exceed that previously experienced, and may exceed the rate at which species can migrate or adapt. It has been estimated that boundaries of climate regions could move poleward at average rates exceeding by a factor of ten the rate at which trees have migrated in the past (Sedjo and Solomon, 1989). Migration of wild species may be further inhibited by cities, roads, and other physical obstacles that result from human development and have created a pattern of "ecological islands" over much of the terrestrial surface (Weiner, 1990).

Effects on Human Activities

Humans can probably adapt to many of the consequences of climatic change. As Schelling (1983) emphasizes, the changes in climate likely to occur in any location are far smaller than the changes many people experience by migration. So long as the changes are moderate, the main effect may simply be the transitional costs of modifying physical infrastructure (e.g., building dikes to protect coastal cities, building new water-supply systems) and developing new crop strains and agricultural practices suited to the changed climate. However, the costs of adaptation and the accompanying social and economic disruption could be large, in part because a large share of the population lives in threatened low-lying areas (e.g., Bangladesh, the Nile delta). The wealthier, more industrialized nations are likely to find adaptation much less onerous than poorer, agricultural nations (Lave and Vickland, 1989). If changes are not moderate, e.g., if general oceanic circulation patterns shift, the consequences could be severe. An issue that has received relatively little attention is that of when global climates would reach a new equilibrium; so long as GHG emissions continue to increase, change may continue (although perhaps at a decreasing rate, as the "atmospheric window" is gradually closed).

From the human perspective, the consequences of climate change are not uniformly adverse. Particularly in the Northern populated regions, higher temperatures would reduce heating expenses and extend agricultural growing seasons (assuming adequate water supplies and suitable soils). In addition, the enhanced concentration of CO_2 may stimulate plant growth by facilitating photosynthesis. Whether this change would on balance favor agricultural commodities or weeds is not known, however (Adams et al., 1990).

Climate change could have adverse effects on human health (Leaf, 1989). For example, the number of abnormally hot days may increase more than in proportion to the mean temperature rise, and could lead to more frequent incidence of heat stress, particularly among the elderly. Industrial air pollution and associated respiratory illness could increase, and tropical disease vectors might find an expanded range of suitable habitat.

Relationship to Stratospheric Ozone Depletion

The issue of global climate change is closely related to that of stratospheric-ozone depletion. Both involve consequences of human-caused emissions of substances to the atmosphere; indeed, the substances of concern to ozone depletion are also contributors to climate change. The two issues interact in that changes in stratospheric-ozone concentrations can affect the radiative balance and climate, and global warming would cool the stratosphere, slowing ozone-loss processes and thereby enhancing ozone concentrations.

Because both issues involve gases with long atmospheric residence times, the consequences of current actions extend well into the future. This makes both issues problems of the global common, since releases from all countries are well mixed in the atmosphere, and have the same environmental effects. In both cases, there is substantial scientific uncertainty about the physical outcomes of past and continuing releases; because the processes are global, it is not possible to conduct definitive experiments that are limited in scope. Consequently, assessments of the likely consequences are derived from complex computer-based models, and the possibility of catastrophe—dramatic ozone reductions or a runaway greenhouse effect—cannot be entirely excluded.

The possibility of climate change differs from potential ozone depletion in important aspects, however. The release of greenhouse gases is much more fundamentally tied to a wide range of human activities, including agriculture, meat and dairy production, and fossil-fuel combustion, the primary energy source since the industrial revolution. Also unlike ozone depletion, the consequences of global warming do not appear entirely negative. Although increased ultraviolet flux at the earth's surface appears to benefit neither humans nor other species, warmer climates and increased CO_2 concentrations could be favorable for some species in some locations. In part because of the apparent difficulty in preventing significant climate change, and the possible benefits, adaptive strategies have received more favorable attention than in the ozone-depletion case.

In the remainder of this chapter, I discuss how the tools described in earlier chapters could be applied to policy analysis relevant to potential global warming. The discussion is divided into three parts: characterizing uncertainty about outcomes, timing responses, and valuing consequences.

CHARACTERIZING UNCERTAINTY ABOUT OUTCOMES

Figure 24 provides a schematic overview of the causes and consequences of global warming. Like the corresponding figure for ozone depletion (Fig. 3), it shows a chain of influences from government policies to eventual consequences; because the causes of GHG emissions and consequences of change are so pervasive, the categories in the figure are somewhat more aggregated than in the ozone-depletion figure.

As for the ozone-depletion case, other factors intervene at every stage: human activities are influenced by population, economic and technical factors, cultural norms and traditions, and other factors in addition to government policies. Greenhouse-gas fluxes, both emissions and withdrawals from the atmosphere, are affected by human consumption, industrial, and agricultural practices, but also by natural processes including some occurring in wetlands, tundra, and forests. These natural processes are in turn influenced by human activities which at a minimum affect the area remaining in a natural state. Overall climate is affected by processes including volcanic emission of aerosols, changes in solar flux, possible biological feedbacks, and other factors as well as the concentration of GHGs. Finally, the influence of climatic change on the natural world is mediated by the ability of species to migrate, human influences including release of other pollutants, and other factors.

An important difference between Fig. 24 and the corresponding figure for ozone depletion is the inclusion of a feedback arrow, reflecting the effects of changes in the natural world on human activities. Changes in climate and in the scope of services provided to humans by natural ecosystems may affect human activities in ways that consequently influence GHG fluxes. For example, changes in ambient temperature may affect energy use for heating and cooling; the difference in energy use may be reflected in changed emissions of CO_2, CH_4, and other GHGs that are produced by fossil-fuel combustion. Similarly, climate change could influence agricultural practices, with corresponding effects on release of N_2O and other GHGs.

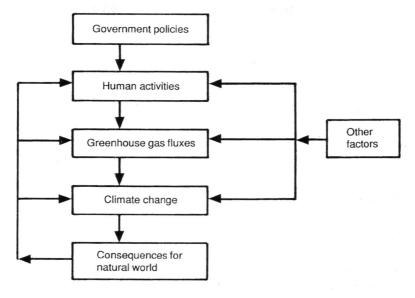

Fig. 24 — Schematic view of potential causes and effects of global warming.

The importance of this feedback link probably increases with the length of the time scale under consideration. If we seek to address only changes in the next decade or two, the feedback may not be very important, because of the long lags between increases in GHG concentrations, climate change, changes in human activities, and consequent changes in GHG fluxes. In the ozone-depletion case, the feedback appears to be of only minor importance, because the consequences of ozone depletion do not appear to significantly affect use and emission of potential ozone depleters. However, the broader linkages between stratospheric-ozone concentrations and climate suggest a feedback effect even in this case. For example, if CFC-11 and 12 emissions are reduced by substituting other compounds as refrigerants and components of insulating foams, an energy penalty may be incurred, possibly increasing CO_2 emissions.

Greenhouse Gas Scenarios

As for the ozone-depletion issue, complex computer-based models have a central role in characterizing the consequences of increases in GHG fluxes. Because these models are so time consuming and expensive to run, complete Monte Carlo analyses of the uncertainties in model structures and inputs are not feasible. Again, a method of choosing scenarios to efficiently characterize the uncertainty about model inputs, including future greenhouse-gas fluxes, appears useful.

The method used for constructing potential-ozone-depleter emission scenarios appears to be useful for constructing similar scenarios to characterize uncertainty about future GHG fluxes. This method uses a score function to reduce the dimensionality of the inputs by constructing scenarios based on the net climatic effect of greenhouse gases. Scenarios of GHG fluxes can be constructed by reference to a probability distribution function for a score function that represents net greenhouse forcing, where the fluxes of each GHG are weighted by their relative greenhouse effects. A limitation of this technique is that it precludes incorporating analysis of the direct effect of enhanced CO_2 concentrations on plant growth. In the previous case, only the combined effect on stratospheric ozone was important; in this case, the concentration of some individual gases may be important, adding additional dimensions to the analysis. One method to incorporate the independent fertilization effect would be to use a two-dimensional score function, with one dimension representing net greenhouse forcing and the other, CO_2 concentration.

The relative contributions of the GHGs to greenhouse forcing depend on several factors. A compound's instantaneous radiative forcing depends on its absorption spectrum and the extent to which it overlaps the absorption spectra of other atmospheric constituents. The instantaneous forcing is subject to a saturation effect, in that successive increments produce less than proportional increases in forcing if the concentration of the compound or others with overlapping absorption bands is sufficiently high. The current atmospheric composition is such that this saturation has begun to occur for CO_2, although not for the other GHGs (Lashof and Ahuja, 1990; Rodhe, 1990).

The cumulative forcing depends on the atmospheric residence time and the pattern of decay. For CFCs and other compounds with only stratospheric sinks, the decay pattern is well represented by an exponential function and is not strongly dependent on the concentrations of other constituents; for CO_2 and CH_4, neither statement holds. Uncertainty about the magnitude of nonatmospheric sinks for CO_2 and their response to increased atmospheric concentrations contributes to uncertainty about the effective residence time for this compound. A further complication is that fluxes of one GHG to the atmosphere can influence the concentrations of other greenhouse gases, so the net effect of emitting a unit of one GHG must account for changes in concentration and greenhouse forcing of other gases. As examples, emissions of CFC-11 and 12 may reduce stratospheric ozone concentrations, and emissions of CH_4 can increase concentrations of tropospheric ozone, water vapor, and CO_2.

Several recent studies have reported estimates of the relative greenhouse forcing associated with GHGs. Fisher et al. (1990b) define the halocarbon global warming potential (HGWP) as the ratio of the calculated steady state infrared forcing of a compound relative to its (constant mass) emission rate, compared to the same ratio for a reference compound (CFC-11). Using two radiative-convective models, HGWPs are calculated for 16 chlorinated compounds including the principal CFCs and their potential replacements in commercial applications. The results differ somewhat between the two radiative-convective models used, but the greenhouse forcings associated with methyl chloroform, HCFC-22, and the HCFCs that are undergoing development and testing as potential substitutes for CFCs are one to two orders of magnitude smaller than the forcings estimated for CFCs 11, 12, and 113. These calculations do not include effects due to changing the abundance of other atmospheric constituents (e.g., ozone) as these were estimated to be of minor importance.

Lashof and Ahuja (1990) and Rodhe (1990) provide analogous results for other GHGs. Lashof and Ahuja define an index of global warming potential (GWP) as the ratio of the time-integrated product of the instantaneous radiative forcing and fraction of the initial emission remaining in the atmosphere to the corresponding ratio for a reference gas, here CO_2. This definition is conceptually equivalent to the HGWP if radiative forcing varies linearly with concentration of a GHG (i.e., no saturation effect) and its atmospheric residence time is constant, conditions that are currently good approximations for CFCs but less so for other GHGs. Rodhe (1990) uses a similar definition, and both authors incorporate the indirect effects of CH_4 breakdown yielding H_2O and CO_2, which Rodhe estimates doubles the net forcing associated with CH_4 emissions. Results of the two studies are of the same order of magnitude, although Rodhe calculates greenhouse forcings (relative to CO_2) two to three times larger than Lashof and Ahuja do for CH_4, N_2O, CFC-11 and CFC-12. These studies find GWPs (per unit mass relative to CO_2) of 10 to 30 for CH_4, 180 to 300 for N_2O, 1,300 to 4,000 for CFC-11, and 3,700 to 8,000 for CFC-12.

Subjective-probability-based scenarios for GHG fluxes could be constructed in a similar manner to those for PODs described earlier. The weights used to construct the score function could be based on the studies just described. For each GHG to be included, subjective probability distribution functions for the net flux could be constructed by analysis of historical emission trends, the specific human activities that influence fluxes, and factors that affect the extent of those activities. Correlations between fluxes of different GHGs are likely to be more significant than for PODs, as some activities produce multiple GHGs and attempts to reduce fluxes of one GHG may increase those of others. For example, fossil-fuel extraction and combustion is a primary source of CO_2 and CH_4. Fluxes of both gases may be positively correlated through their relationship to total energy use, or negatively correlated if natural gas is substituted for other fossil fuels. Similarly, any increase in energy use associated with the diminished use of CFCs for insulation or refrigerants could produce greater CO_2 fluxes.

Unlike the POD-emission scenarios, GHG-flux scenarios would need to account for possible changes in non-atmospheric sinks, e.g., potentially reduced biospheric uptake of CO_2 because of deforestation. It may be most useful to construct separate distribution functions for individual sources and sinks of each gas, and aggregate them to obtain probability distributions for net fluxes of each gas. This approach would parallel the analysis of individual POD applications undertaken as foundation for the POD scenarios.

Understanding of the strength of future sources and sinks of individual GHGs varies widely. Considerable work has been done in projecting future CO_2 emissions from energy use (e.g., Edmonds et al., 1984; Nordhaus and Yohe, 1983), although other sources and sinks are less well understood (Tans et al., 1990). CFC and other potential ozone depleter emissions have been well studied, although these analyses generally do not account for the effects of the Montreal Protocol, the international agreement to reduce POD emissions. Wigley (1988) and Office of Technology Assessment (1988) assess likely emissions under the Montreal Protocol as originally ratified, but at the time of writing (summer 1990) it appears likely that prominent nonsignatories (China and India) will ratify the Protocol, and that substantially more stringent provisions will be adopted. In addition, expanded research into POD substitutes that attended the international agreements may significantly alter the course of POD emissions relative to that projected earlier (Manzer, 1990, describes progress in making substitute compounds commercially available). As noted earlier, understanding of the current fluxes of CH_4 and N_2O is limited, and projections of future concentrations have been limited to extrapolation of observed trends.

TIMING RESPONSES

The fundamental problem of deciding when to take action to prevent, or at least reduce, global warming is closely similar to the problem of when to reduce CFC and other emissions to prevent or reduce ozone depletion. Although global warming is more complex, the machinery introduced in the ozone depletion context also appears to be applicable and useful to global warming.

The essential conceptual components of the analysis appear to be directly transferable. In global warming, as in ozone depletion, the pollutants remain in the atmosphere for decades and centuries. Reducing net fluxes would also require changes in human activities that may require years to accomplish. Annual variations in net fluxes are not important; only the cumulative flux over longer periods is relevant. As described in the previous section, weights analogous to those for relative ozone-depletion efficiencies can be derived to represent the relative contributions of GHGs to global warming. Consequently, it appears that a constraint on cumulative weighted fluxes through some planning horizon can adequately account for the environmental forcing of greenhouse gases, and policy analysis can be structured in terms of limiting that quantity to various levels.

In principal, a cost function for reducing GHG fluxes can be constructed, analogous to the cost function for reducing POD emissions. Such a cost function should incorporate measures to increase the non-atmospheric sink strengths (e.g., by reforestation), as well as reducing source strengths. Because many of the alternatives available for reducing GHG fluxes involve fundamental changes in human activities, estimating the costs of these actions may be more difficult than in the ozone-depletion case, where most of the steps involved input substitution in industrial processes. Some of these difficulties and measures for dealing with them are described in the section on valuation below. The feasible technical alternatives for reducing GHG fluxes and their costs may also vary widely between nations, particularly those at different stages of economic development (Crosson, 1989; Jodha, 1989; Lave and Vickland, 1989).

Estimates of the costs of reducing GHG fluxes in future years become increasingly uncertain because of the potential role of new technologies. Energy technologies that would produce few or no greenhouse gases include photoelectrics, current or later generation nuclear, and use of hydrogen as a transportation fuel. Total energy demand might be substantially reduced by conservation (Lovins, 1976; Lovins et al., 1981; Goldemberg et al., 1987; Ausubel and Sladovich, 1989; Oppenheimer and Boyle, 1990).

In the ozone-depletion case, the results of the timing analysis were characterized by a critical probability, which is a function of the acceptable cumulative weighted emissions. This probability is intended to be compared to another probability that summarizes a decision maker's assessment of the likelihood of sufficient ozone depletion occurring to have substantial adverse consequences. This second probability is the probability that advances in understanding the science and consequences of ozone depletion will reveal that curtailing emissions to remain within the cumulative emission limit will be desirable. The critical probability is defined such that a decision maker who estimates the probability of learning that emission curtailments will be desirable as exceeding the critical probability should prefer to begin reducing emissions as soon as possible; if he estimates the chance of learning that curtailments are desirable as smaller than the critical probability, he should prefer to wait and then reduce emissions only if later information shows that to be appropriate.

In the global-warming case, uncertainty about the value of the critical probability for any specified cumulative emission limit is likely to be large, because of uncertainties about the current and future cost functions for reducing GHG fluxes. Uncertainty about the appropriate cumulative emission level is also substantial, as it depends on the sensitivity of the climate to incremental greenhouse forcing, and on the relative costs of preventing or adapting to climate change. Even if the critical probability is only moderately sensitive to the specified emission limit, this source may also contribute significantly to uncertainty about the critical probability.

Estimating the Costs and Benefits of Delaying Action

Even if uncertainties about its value preclude using a critical probability for policy choice, this framework and the research tasks taken to implement it can provide valuable information for clarifying policy decisions. For example, a first step in this analysis could be to identify and characterize the benefits and costs of delaying activities that would reduce GHG fluxes. The costs and benefits could be characterized for several lengths of delay, e.g., two, five, and ten years.

There are two principal costs to delaying actions that reduce GHG fluxes. One is the cost imposed by increasing the cumulative atmospheric burden of GHGs. Because the atmospheric residence times of the GHGs are on the order of decades or even centuries, current fluxes effectively preclude the possibility of stabilizing GHG concentrations at relatively low levels. Moreover, because of thermal inertia of the oceans and cryosphere, the current climate lags behind the equilibrium climate associated with the current atmospheric composition. These factors are recognized by the concept of "warming commitment" (Mintzer, 1987), the equilibrium increase in global average temperature associated with the current increase in GHG concentrations relative to pre-industrial levels. Additional GHG fluxes effectively commit the earth to a warmer climate for the foreseeable future.

The cost of delaying action for a specified period can be quantified in terms of the increased atmospheric burden of GHGs, because current accumulation rates are known with relatively high accuracy, and can be projected into the near future under a "business as usual" scenario. The projected increase in GHG concentrations can also be scaled to the cumulative increase since pre-industrial times, allowing comparisons of the incremental effect of delay relative to the total effect of GHG increases. The actual warming commitment and associated change in regional climates associated with the increased atmospheric burden cannot, of course, be reliably estimated at present.

The second cost of delaying actions to reduce GHG fluxes is the potentially higher cost of limiting cumulative fluxes to any chosen level. By delaying actions to reduce fluxes, the opportunity to use relatively cost-effective measures in early years is precluded; fluxes that could have been eliminated by relatively inexpensive steps in early years may need to be offset by more expensive steps in later years. For example, cumulative CO_2 emissions from automobiles could be reduced a certain amount if the average fuel efficiency of new vehicles improved five miles per gallon beginning with the next model year; if improvements are delayed five years, a larger and presumably more costly efficiency improvement would be needed to obtain the same cumulative emission target.

The second cost can also be quantified, to some extent. To do so requires identification of promising actions to reduce GHG fluxes, and assessment of their cost-effectiveness in reducing greenhouse forcing. Incremental greenhouse forcing may be approximated as the product of the incremental flux and the relative global warming potential described above. Ranking these actions by cost effectiveness provides a method of estimating the increased cost of offsetting near-term fluxes by a more aggressive portfolio of actions later. As noted earlier, these estimates are subject to uncertainties related to the rate at which new technologies can be developed and integrated into economic systems and to the costs associated with these technologies.

The benefits associated with delaying any decision to adopt flux-limiting policies involve anticipated improvements in understanding the physical science of climate change and the policies available to limit change or adapt to it. Improved information should allow better decisions in that the most cost-effective measures can be identified and inappropriate measures can be avoided. Predicting what will be learned is always difficult, but several methods may provide information about the likelihood that what is learned will be sufficient to affect policy choice. If the improved information that can be obtained by delaying action is not likely to influence decisions about whether to act and what measures to adopt, that information has no value for policy making (Hammitt et al., 1989; Hammitt and Cave, 1990).

One method to assess the likely benefits of improved information is to evaluate whether a few more years of climate observations can provide convincing evidence to support or refute the results of current model simulations. An important area of scientific disagreement concerns the relationship between accumulated GHGs and observed climate change to date. Historical temperature records suggest the global average temperature has increased perhaps 0.5° C over the last century, which is near the lower end of the model-simulated increase (Jones et al., 1986, 1988; Hansen and Lebedeff, 1987; Schneider, 1989). The observed pattern of temperature increase has not been monotonic and accelerating in parallel with GHG concentrations; most of the increase occurred between about 1910 and 1940, and during the 1980s. Satellite data available only for the last decade, however, reveal substantial variability in temperature, but no clear trend in the 1980s (Spencer and Christy, 1990). In other periods the global average temperature apparently held constant or even declined. Kuo et al. (1990) find a high degree of coherence between atmospheric CO_2 and global average temperature over the past 30 years, but Barnett (1990) cautions against over-interpretation of this result.

The reasons for the discrepancies between observed temperatures and model simulations are unknown, but may include natural variability in the climate occurring on decadal and longer time scales (e.g., El Nino-Southern Oscillation events), other climate forcings (e.g., volcanic aerosols, changes in solar luminosity), and measurement or sampling error in the temperature record. To the extent possible, the global temperature average has been corrected for errors related to movement of thermometers and the growth of urban heat islands around them, changes in measuring procedures and technologies, and uneven geographic sampling. However, uncertainties remain about the proper corrections and possible bias related to non-uniform sampling. In particular, measurements over the oceans and polar regions are relatively sparse and geographically concentrated along trade routes.

A first step in analyzing the value of additional observations would be to statistically evaluate the existing records, and to calculate the degree to which several additional annual observations would narrow the confidence region for the estimated trend. It would also be possible to estimate the probabilities of observing temperature series over the next several years that would allow statistical discrimination between hypotheses based on model simulations and a null hypotheses of no trend. An important issue to address would be the degree of independence between annual observations, and the extent to which lack of independence degrades the value of additional observations. Such an analysis could be improved by accounting for variations in the quality of observations, where new observations can be made much more reliable than those reconstructed from historical data. An example of analysis of this type is reported by Hansen et al. (1988), who estimate that a 0.4° C increase above a 1951-1980 reference period would be statistically significant evidence of warming, and report that model simulations strongly suggest this increase will be observed during the 1990s.

This approach of estimating the incremental value of additional observations of some climate statistic can be applied to other statistics as well, including the annual temperatures at various heights in the atmosphere or depths in the seas, regional or seasonal temperature or precipitation averages, or frequencies of extreme events. An important issue is to determine which climate statistics are likely to be most informative. Important factors in this evaluation include the magnitude of change expected relative to natural variability (i.e., the signal-to-noise ratio), the rate at which the statistic is expected to respond to climate change, the quality of historical records, and the cost and feasibility of ongoing monitoring. Some combination of climate statistics (e.g., a global-warming "fingerprint") may provide the most powerful indicator of climate change (Weller et al., 1983).

Based on climate simulations, Hansen et al. (1988) suggest that summer temperatures for middle and low latitudes may exhibit relatively high signal-to-noise ratios, although not as high as the global average temperature. The lower climate noise in the global average might be offset by greater measurement variation, however. Analysis of climate statistics suggests that stratospheric temperatures, atmospheric radiation, sea level, sea-ice and snow cover may be useful indicators (Weller et al., 1983). Some thinning of the Arctic ice pack has been reported, although it cannot be distinguished from natural variability because of limited baseline data (McLaren et al., 1990; Wadhams, 1990).

A second approach to estimating the value of delay might be to attempt to predict advances in scientific understanding and modeling capabilities. For example, observational studies or laboratory experiments may be undertaken that provide information to refine understanding of critical components of the climate system; e.g., the Earth Radiation Budget Experiment is providing information about the role of clouds (Ramanathan et al., 1989). Potential improvements in the resolution of climate models or the linking of atmospheric and oceanic GCMs might be projected from an assessment of the likely increases in computing resources; however, required computing resources are a geometric function of the number of grid points, and the value of better resolution is unclear. The value may be small unless resolution is improved to a level that allows more detailed treatment of subgrid-scale processes.

The discussion so far has addressed the issue of how to confirm if global warming is occurring, with a view towards implementing preventive measures if it is. Policies designed to adapt to changes in climate (other than sea-level rise) are likely to be more sensitive to regional climate changes than to global averages. For deciding when to begin planning or implementing adaptive measures (e.g., modifying water-resource systems, agricultural or construction practices), detecting the magnitude and sign of changes in regional climate variables is more relevant. Because the signal-to-noise ratio for small-area changes is likely to be much smaller than for well-selected global changes, it may be more difficult to identify changes on regional than global scales, however. It is unclear whether regional changes can be most accurately forecast from direct observation or by using GCM simulation results combined with observation of global climate statistics.

Implementation Delays

Many of the potential measures to reduce GHG fluxes entail significant delay from the time a decision is made to adopt it until the measure becomes effective. Adaptive measures may also require substantial implementation periods. These lags may result from the time required to design and plan facilities or to develop new industrial technologies, crop strains, or agricultural practices; to the rate at which existing infrastructure is retired; and to other factors. In addition to these primarily technical factors, the time needed to reach agreement among national governments and other parties to undertake a coordinated response may be an important factor.

Characterization of the lags between a decision to adopt a measure and when it achieves various stages of effectiveness, and how those lags can be reduced with increased resources, would be useful in determining when to take actions. Comparison of the time required to implement adaptive measures with the rate at which climate may change will suggest whether certain adaptive measures should be initiated before it becomes certain they will be needed. Assessment of implementation delays should also be integrated with the estimated costs of delaying action, as the warming commitment will increase because of these delays. In general, planning activities may be relatively attractive for the near term, as they may reduce the lags attendant on adopting particular measures. But the limited accuracy with which the magnitude, and even the sign, of changes in climate variables can be predicted on regional scales constrains the planning that can be undertaken. Sets of plans prepared for multiple contingencies may be most cost effective.

Information about implementation lags provides insight into which measures should be adopted first, if any action is taken. Consideration of these lags suggests two candidates for early action: preventive or adaptive measures with long lags between adoption and effect, and preventive measures that become effective relatively quickly. Measures in the first category merit attention as they may not become effective in time to have much benefit if initiation is delayed; i.e., if sea walls are needed, they should be built before major storms cause significant damage to the facilities they are designed to protect. Similarly, measures to reduce concentrations of CO_2 and PODs must be implemented long before they produce much reduction in the atmospheric concentrations of these gases, because of the probable long delays in replacing or adapting existing capital and the long atmospheric residence times of these GHGs.

Measures in the second category, preventive actions that become effective relatively quickly, may provide useful insurance by slowing global warming (if it is occurring), thereby providing additional time to understand the causes and consequences of warming and to begin other preventive or adaptive measures. Whether such measures are appropriate depends on how quickly warming could be halted or accommodated when its existence becomes more certain. Many preventive measures have long delays between implementation and effect, but measures to influence CH_4 fluxes could be attractive. Because the atmospheric residence time for CH_4 is believed to be an order of magnitude shorter than that of other GHGs (about 10 to 15 years; Lashof and Ahuja, 1990; Rodhe, 1990), atmospheric concentrations are more sensitive to current fluxes and the rate of concentration increase may be more rapidly slowed (depending on the effectiveness of available measures in reducing fluxes). To date, relatively little attention has been directed towards identifying CH_4 control measures beyond reducing unintended emissions from the extraction and distribution systems, however.

VALUING CONSEQUENCES

Valuing the potential consequences of global warming is complex. The diversity of consequences suggests that no single technique will be sufficient, and a variety of methods may contribute. Because many significant consequences may not be realized until relatively far in the future, valuation is especially sensitive to issues involving intertemporal, and even intergenerational, choice. Moreover, the consequences are likely to vary substantially by locality or region (e.g., Lave and Vickland, 1989; Edgerton et al., 1990; Montgomery, 1990) and to include some benefits which must be offset against costs. The pervasive nature of consequences may require accounting for interactions throughout the economy, a general-rather than partial-equilibrium analysis (Kokoski and Smith, 1987).

In order to assign values, it is necessary to be precise about the consequences being evaluated. A useful first step is to value the difference in economic consequences between the present economy and a hypothetical one adapted to some benchmark, e.g., the simulated equilibrium climate corresponding to a doubled CO_2 greenhouse forcing relative to pre-industrial times (e.g., Adams et al., 1990, estimate this difference for U.S. agriculture). Although useful, this steady-state comparison is incomplete as it neglects a number of important factors, including economic values associated with transient conditions, uncertainty about the appropriate equilibrium climates, and the effects of technological developments on adaptation to climate change.

Comparison of two economic/climate equilibria is misleading, because of the long lags in the climate system. Whenever the increase in GHG concentrations stops, the climate may not reach equilibrium until decades later. Similarly, the time needed to stabilize GHG concentrations and for the economy to adapt to the new climate virtually ensure that many decades will be spent in non-equilibrium conditions. The economic value of the transient effects may dominate the difference between the prior and subsequent equilibria, especially if future consequences are discounted at a significant rate. Although transition costs are frequently important in policy analysis, their measurement has received relatively little attention; some attempts to develop applicable methodology have been reported, however (e.g., Camm and Kohler, 1987; Cave, 1988).

The eventual climate equilibrium that will be reached if global warming occurs is also uncertain. The widely used assumption of a doubled CO_2 concentration is only a convenient benchmark; there is no assurance that GHG concentrations will not exceed this level. Preliminary and unpublished results suggest the costs to the United States or other countries of limiting GHG fluxes to levels that might begin to stabilize concentrations at the radiative equivalent of doubled CO_2 may be several percent of gross domestic product or more (e.g., Manne and Richels, 1989, 1990; Nordhaus, 1989). The economic consequences associated with this equilibrium cannot be portrayed as the continuation of current activities under different climatic conditions. Rather, the economic consequences must include consideration of the appropriate degree of adaptation, including relocation of climate-influenced activities to more suitable sites, changes in crops and agricultural practices, and other changes in economic activities.

The base point for the comparison—the "current" climate—may not be relevant for policy analysis. The global average temperature appears to have increased about 0.5° C over the last century (Jones et al., 1986; Hansen and Lebedeff, 1987). Even if GHG concentrations were stabilized at the current level, the current climate may not yet have equilibrated to the existing greenhouse forcing and could continue to shift for some time.

Because the consequences of global warming are expected to be long-lived, significant technological development is likely to occur during this period and to influence the economic costs and benefits. The long-term costs associated with a different climate may be substantially mitigated by the development of technologies better suited to that climate. Technological development might include new crop strains and agricultural practices, for example. Because the resources invested in developing technologies adapted to climate change are not available for other uses, these costs are part of the costs of climate change.

As this discussion suggests, policy analysis may be most useful if it identifies alternative feasible policies and projects the associated trajectories of consequences through time. The economic values of differences in consequences between these trajectories, at each future date, are the relevant values for policy choice. To account for uncertainty about the outcomes of each candidate policy, it may be useful to construct several scenarios (internally consistent trajectories) for each policy. The expected value of the consequences corresponding to a policy can be estimated by combining the values of consequences along each scenario with reference to the probability associated with that scenario. Clearly, construction of regionally detailed scenarios for alternative policies is highly speculative at present, but policy analysis that explicitly addresses these uncertainties may be more useful than analysis that ignores them.

Methods for Valuing Consequences of Global Warming

The consequences of global warming may be divided into classes based on the methodologies available for valuing them. In the first class are consequences for which engineering-cost methods capture a major component of the costs and benefits, and may be adequate. Engineering-cost methods attempt to identify the resources needed to complete a project, e.g., quantities and types of workers, raw materials and tools, and estimate the total cost as the sum of products of input requirements and market- or cost-based prices. Adaptive measures such as construction of seawalls, irrigation, and water-supply systems may be adequately valued by estimates of design and construction costs. Similarly, depletion of natural resources such as timber or commercially harvested fish may be valued by market prices or the costs of managed production or substitutes. The costs of measures that require substantial technological innovations, such as more heat-tolerant crop strains, are more difficult to estimate because of the inherent uncertainty in predicting the difficulty of developing appropriate technologies, but here also the engineering-cost perspective appears useful. In some cases, the broader environmental consequences of these projects will also be significant and may require other methods of analysis (e.g., the effects on ecological communities, recreational and aesthetic opportunities that may attend flooding a canyon to create a new reservoir).

Even when engineering-cost methods appear adequate, a broader economic analysis may provide important refinements. The effectiveness of installing more conservative physical capital (e.g., energy-efficient heating and cooling systems, improved building insulation, low-flow shower heads and toilets) may not be adequately captured by analysis that does not allow for possible changes in behavior. If better insulation reduces the marginal cost of heating or cooling, part of the anticipated energy savings may be offset as individuals reset their thermostats to more comfortable temperatures (Friedman, 1987; Dewees and Wilson, 1990). Seawalls and other structures intended to protect low-lying areas from storm surges may attract greater economic investment to the area, which will be at risk if the seawall is breached and may have deleterious environmental consequences.[4] Drawing from another field, some evidence suggests that government-mandated improvements in automobile safety equipment were not as effective as anticipated, because motorists drove less cautiously once ensconced in more occupant-protective vehicles. Although automobile-occupant fatalities were somewhat lower than before introduction of the new equipment, the decrease was smaller than expected and partially offset by increased pedestrian mortality (Peltzman, 1975a, 1975b; cf. Graham and Garber, 1984; Asch, 1988).

A second class of consequences cannot readily be valued using the engineering-cost approach. These include changes in natural ecosystems, extinction of particular species or a general decline in biodiversity, and others. These values may be partially accounted for by the concepts of use, option, and existence values.

Use and option values incorporate the instrumental role of environmental attributes in satisfying human desires. Use values refer to the benefit of environmental services provided. For example, the aesthetic and recreational benefits of natural settings have been estimated by revealed-preference (e.g., travel cost) and contingent-valuation methods. Similarly, part of the value of perturbations to selected ecosystems might be captured by assessing changes in capacities for providing environmental services (e.g., waste assimilation and water purification by wetlands, CO_2 uptake by forests). It may be important to evaluate current or potential benefits provided through small scale, decentralized economic activities that may not be adequately captured by national accounting statistics; for example, the value of sustainable yields of "minor" rainforest products (e.g., fruits, oils, latex, medicines) may dwarf the value of timber or agricultural use (Peters et al., 1989).

Option value incorporates the value of uncertain use; i.e., even though a resource is not used at present (or not used by an individual), there may be value in maintaining the resource in case its services are desired in the future. The value of biodiversity or of large undisturbed ecosystems may be characterized in part by the new products or knowledge that may be derived from them (Norton, 1986). The existing biota may be viewed as a library of biological and chemical arrangements that may be used directly, or with adaptation, for human products such as medicines (it is estimated that 25 percent of Western drugs originated from substances isolated from rain forests; Mattos de Lemos, 1990). Such option values might be estimated by conventional economic methods incorporating estimates of the probability of use occurring; e.g., environmental services or new products may be valued by the reduction in cost relative to alternative means of providing these services. Contingent-valuation methods may also be applicable.

Existence value differs from these instrumental values, in that it does not assume any direct use of the resource will ever occur. People may derive satisfaction from the knowledge that certain species or ecosystems exist; this satisfaction may be based on the idea that humankind is not entirely free to use other elements of the biosphere for its own wants, but should recognize and respect the natural order of things (Stone, 1972; Tribe, 1974; McKibben, 1989). To the extent that such existence values are incorporated in individual utility functions, they may be estimated using a contingent-valuation approach.

Challenges to Welfare Economics

Attempts to value some of the possible consequences of global warming highlight limitations of economic welfare theory and the utilitarianism from which it derives. The long time lags inherent in this issue require making intertemporal and even intergenerational choices, emphasizing the choice of discount rate. Because the individuals alive now and those who will be alive when many of the consequences of global warming may be felt are different, intertemporal choices on this time scale are inherently interpersonal. Welfare economics provides little guidance on this issue. The principal recommendation of welfare economics is to suggest that, if future generations are more wealthy, their marginal utility of income may be smaller than the current generation's, and so current sacrifices to improve future incomes should be limited. If future generations are wealthier, however, they may also assign higher values to human health and environmental quality than the present generation, both in absolute willingness to pay and relative to other goods and services.

In a cost-benefit analysis of stratospheric-ozone depletion, the U.S. Environmental Protection Agency assumed the value of human life would increase over time, reflecting increases in real wealth. This increase partially offset the effect of discounting on the relative weight given to current and future consequences. In the "medium" or reference case, EPA discounted future values at two percent (real) but assumed the value of life would increase at 1.7 percent annually, largely offsetting the effect of discounting on this component. In its "high" scenario, the value of life was assumed to increase at 3.4 percent annually, overwhelming the one percent annual discount rate used in this case (EPA, 1987).

Discounting is best justified by the argument that a quantum of resources invested today can be expected to yield a larger quantity in the future, so the opportunity cost of spending the resources now is foregoing the option of spending a larger quantity in the future. When the future benefit foregone is comparable to the present benefit, this rationale is compelling, but when the future benefit goes to another population, or when irreversibilities intervene so present opportunities foregone are not available in the future, the rationale is less persuasive. If current opportunities to limit global warming or preserve endangered species are not exercised, future generations' greater wealth will not allow these opportunities to be reclaimed.

To the extent that utility functions are determined by social and cultural norms that may change with time, relative values of consequences may shift in directions that are difficult to forecast. Western aesthetic standards have shifted with fashion and perhaps with more fundamental factors; today, a well-tanned and athletic physique is desired, but this was not true when such attributes distinguished workers from elite classes. The personal transportation services and self-image provided by automobiles are today highly valued in large segments of western society, but may become marks of indecent consumption in the future. Douglas and Wildavsky (1982) explore the processes by which cultures choose the risks to which they direct greatest attention.

The values of consequences associated with alternative policies are likely to be sensitive to regional, seasonal, and temporal details of climatic change. Because the detailed consequences associated with any particular policy are subject to broad uncertainties, refined estimates of the values associated with particular consequences may not be important, at present. A more useful approach may be to attempt to construct gross bounds on the values associated with particular consequences, in order to provide some metric with which the costs of preventive actions may be compared. To the extent that both humans and nature can be expected to adapt to moderate changes in climate, the most significant values may be those associated with disruption occasioned by rapid change.

CONCLUSIONS

The possibility of widespread climate change induced by atmospheric pollutants of anthropogenic origin epitomizes many of the challenges inherent in environmental policy choice. Pervasive uncertainty accompanies analysis of each of the links from policy choice to ultimate consequences and their significance. The long time lags between changes in policy and ultimate consequences virtually ensure that some degree of uncertainty about consequences will always remain, yet these lags also require that actions to prevent or mitigate the effects of global warming must begin long before their benefits, if any, will accrue. We can also expect to learn more about the linkages, so that decisions delayed may be better informed and more likely to select well-targetted policies. The tension between the need to act early to prevent what may otherwise become inevitable, and the need to better understand the climate system and other links between policies and outcomes to avoid choosing overly costly or ineffective measures, is the heart of the dilemma. The policy issue is further complicated by difficulties in assessing the significance of, and formally assigning values to, pervasive climate changes that may affect a wide range of human and non-human activities.

I have attempted to characterize some of the fundamental problems inherent in many environmental policy issues, which I broadly describe as outcome and value uncertainties. I have also presented several tools that may be helpful in thinking through policy questions and achieving some resolution. These tools include probability-based scenarios for characterizing uncertainty about factors that may affect environmental consequences, a method to assess the implications of undertaking preventive actions soon or postponing a decision in order to incorporate expected improvements in understanding the consequences of alternative policies, and revealed-preference and contingent-valuation methods for estimating the economic values associated with environmental consequences, which are needed to compare costs of mitigating measures with the costs of simply accepting change.

The fundamental and pervasive uncertainties about outcomes and values, as well as the time lags and potential for learning inherent in many environmental policy questions, suggest that these questions cannot be approached as one-time problems for which a solution can be found and implemented. A more effective and beneficial approach is likely to involve recognition of the ongoing nature of human interaction with the environment and continuing needs to assess and balance the risks of alternative policies, with the understanding that scientific and economic issues will be better understood in the future and that flexible policies that can be adapted as understanding grows may prove best.

NOTES

Chapter 1
Outcome and Value Uncertainty in Environmental Policy

[1]This framework was developed prior to negotiation of the "Montreal Protocol on Substances that Deplete Stratospheric Ozone" (Crawford, 1987). The framework and its application were presented at one of the United Nations Environment Programme/U.S. Environmental Protection Agency-sponsored workshops convened as part of the negotiation process (Hammitt, 1986). The framework remains applicable to the periodic reassessment of terms required under the Protocol, although the data presented in this application require updating for current use.

Chapter 2
The Policy Problem

[1]This Part is based on materials developed and written in 1986 and 1987, prior to negotiation of the Montreal Protocol on Substances that Deplete Stratospheric Ozone. Its perspective is appropriate to that time. The overall framework and methods remain applicable to the periodic reassessments of the Protocol called for in one of its provisions, although the data presented would need to be updated to account for rapid progress in international negotiations and scientific understanding in the last few years. The first articles recognizing the possibility of CFC-induced ozone depletion were Molina and Rowland (1974), Cicerone et al. (1974), and Lovelock (1974). For more information on the current understanding of ozone depletion, the contribution of potential ozone depleters to global warming, and the likely consequences for the biosphere, see the comprehensive three-volume World Meteorological Organization (1985), National Academy of Sciences (1976, 1979, 1982, 1984), National Research Council (1983), Stolarski (1982, 1988), Seidel and Keyes (1983), Wuebbles (1983), Prather et al. (1984), Ramanathan et al. (1985), Connell and Wuebbles (1986), Dickinson and Cicerone (1986), Titus (1986a, 1986b, 1986c, 1986d), Watson et al. (1986), Cicerone (1987), Hammitt et al. (1987), U.S. Environmental Protection Agency (December 1987a, 1987b), Brasseur and Hitchman (1988), Frederick and Snell (1988), Grant (1988), Kiehl et al. (1988), MacDonald (1988), and Schaefer (1988), Taylor et al. (1988), Ramanathan (1988), Vupputuri (1988), and Blumthaler and Ambach (1990).

[2]Annual production and use of the chemicals within regions are treated as equivalent, since inventories and net regional imports and exports are generally small. The difference between production and use in a single nation may be significant, but systematic import and export statistics for each chemical needed to estimate the size of the difference are apparently not publicly available (Hammitt et al., 1986).

[3]Khalil and Rasmussen (1986) measure emissions from rigid foams that occur at a rate consistent with a half life of 100 years or more.

[4]New scenarios reflecting the terms of the Montreal Protocol on Substances that Deplete Stratospheric Ozone could be developed using similar methods. Significant uncertainties to be reflected in such scenarios include whether countries that have not yet joined the Protocol will, the extent to which signatories will comply with the Protocol's provisions, and whether the provisions will be modified in the periodic reassessments called for under the Protocol.

Chapter 3
Probability-Based Scenarios

[1]Subjective probability methods have been used to characterize uncertainties associated with acid deposition (Morgan et al., 1984; Peck and Richels, 1987) and with the effects of air pollution on crop growth (Adams et al, 1984).

[2]This is a slight oversimplification. Certain factors that affect emissions in the top part of the figure may also be important to the size of effects in the bottom half of the figure. For example, the level of general economic activity could affect the size of crops or quantities of materials that might be harmed by ozone depletion. These kinds of dependencies can be integrated by making the distribution for ozone depletion explicitly a function of general economic growth.

Chapter 4
The Distribution of the Score Function

[1]Equivalently, assume that the future quantities of PODs are distributed approximately log-normally. The correspondence between a normally distributed growth rate and a log-normally distributed future production is good for modest growth rates. Let $Y_t = Y_0 (1 + r)^t$ where Y_i is production in year i and r is the growth rate. Then $\log Y_t = \log Y_0 + t \log (1 + r) \approx \log Y_0 + t\, r$ for small r.

[2]This is an approximation that works well for small growth rates. The exact relationship is $(1 + r_k) = (1 + g)(1 + u_k) = 1 + g + u_k + gu_k$, where r_k is the growth rate for the kth chemical, g is the growth rate for GDP, and u_k is the growth rate for the intensity of use of the kth chemical relative to GDP. For small g and u_k, $gu_k \approx 0$ and $r_k \approx g + u_k$.

[3]Alternatives would include using the weights corresponding to the end of the period (a Paasche index) or an intermediate set of weights. As long as the correct weights do not shift much over the period any of these choices will produce similar results. This is a specific example of the general problem of defining index numbers. For further discussion, see Hirshleifer (1976) or other economics texts.

[4]Note that q_c need not be determined to calculate the growth rates relevant to any scenario. Once $z(q_c)$ is calculated from equation (9), it can be substituted into (8) and the appropriate values of the component growth rates can be calculated.

[5]To see this, square the expression that defines β in equation (9) and subtract the denominator from the numerator. This yields an expression

$$2 \Sigma_i [w_i v_g^{0.5} v_i^{0.5}] + 2 \Sigma_{i>j} [w_i w_j ((v_i v_j)^{0.5} - v_{ij})].$$

The first term must be positive. To sign the second, note that the subjective analog of the Pearson correlation coefficient, $v_{ij}/(v_i v_j)^{0.5}$, cannot exceed unity. Hence, the second term is also positive. Therefore the numerator must exceed the denominator and β must exceed one.

Chapter 5
Subjective Marginal Probability Distributions
for Potential Ozone Depleters

[1]The mean of a set of identically distributed variables has a lower variance than the variables themselves and decreases with the number of variables, as long as the variables are not perfectly correlated. Growth rates in different years are expected to be positively correlated, but not perfectly.

[2]Heuristically, one can think of a growth rate as having two components. The first is a secular component that can be represented by a single random variate for each chemical. The second is an annual component that requires separate random variates for each year. These can be independent of one another over time (though not necessarily across chemicals). The first component embodies the positive correlation over time; the second is an additional source of uncertainty that helps explain why the variance of a mean growth rate is smaller in longer periods.

[3]The total growth rate for the ith chemical, r_i, equals $g + u_i$ where g is the general growth rate and u_i is the growth of intensity of use of the ith chemical relative to general economic growth. Hence, $cov(r_i,r_j) = cov(g + u_i,g + u_j) = var(g) + cov(g,u_i) + cov(g,u_j) + cov(u_i,u_j) = var(g)$, since all the covariances in the last step are equal to zero.

[4]Personal communication with Gary Yohe, Wesleyan University, Middletown, Connecticut, 8 January 1986. See also Nordhaus and Yohe (1983).

[5]Global labor productivity and population may be inversely correlated over this period because population increases are likely to be in the poorer nations that suffer from shortages of human and physical capital. Using an input-output model of the world economy, Leontief (1979) simulates how variations in population are likely to be inversely correlated with per capita GDP.

[6]The adjustment term is (2)(-0.25)(standard deviation for population growth rate)(standard deviation for growth rate of labor productivity). This term is small enough that the choice of correlation coefficient does not affect the results much.

Chapter 6
Production and Emission Scenarios

[1]The largest difference is about 0.04, for Halon 1301. Weights corresponding to the median growth scenario in 2040 are CFC-11, 0.262; CFC-12, 0.330; carbon tetrachloride, 0.051; CFC-113, 0.151; methyl chloroform, 0.045; Halon 1301, 0.073; Halon 1211, 0.089.

[2]Gamlen et al. (1986) describe a slightly different algorithm for simulating CFC-11 and 12 emissions used by the Chemical Manufacturers Association in its annual reports. The CMA algorithm assumes slightly more rapid emissions from refrigeration and rigid foam applications.

Chapter 8
Timing Responses to Potential Stratospheric Ozone Depletion

[1]As noted in Section A, this material was written from the perspective of 1986 and 1987. The current (1990) question is whether to strengthen (or relax) the terms of the Montreal Protocol on Substances that Deplete the Ozone Layer or to change other policies that influence POD emissions.

[2]Ehrlich and Becker (1972) distinguish between insurance and protection. Protection reduces the probability of an adverse outcome; insurance reduces the severity of the outcome. This distinction is comparable to that between preventive and adaptive strategies for responding to potential environmental perturbations.

Chapter 9
The Decision Framework

[1]Bayes' rule is a method for updating beliefs expressed as a probability distribution. If the value of some outcome variable z_t is observed and the distribution for z_t conditional on x is known to be $g(z_t|x)$, then $f_{t+1}(x|z_t) = f_t(x) g(z_t|x)/\int f_t(x) g(z_t|x)\,dx$.

[2]Camm and Kohler (1987) analyze the transition costs of regulations and illustrate their methods with an application to the 1978 U.S. ban on most CFC aerosol applications.

[3]Evaluating alternative policies by their expected economic costs alone may be inadequate, particularly when the potential outcomes are far apart. There is a presumption that most people are risk averse when facing potentially large losses; that is, they value a risky alternative less than its expected value (Raiffa, 1968). This is one justification for purchasing insurance. Similarly, risk aversion may be appropriate for evaluating public goods, such as the risk of ozone depletion, that cannot be diversified by society. Risk aversion would tend to make the immediate regulation policy relatively more favorable, thereby reducing the critical probability.

[4]The weighting factors approximate the relative depletion efficiencies of each POD. (The actual depletion efficiencies can vary with the quantities of each POD and of other trace gases in the atmosphere.) The factors used are the same as those used in developing the scenarios to describe possible future depletion above: CFC-11, CFC-12, CFC-113, and carbon tetrachloride, 1.0; methyl chloroform, 0.1; Halon-1211 and Halon-1301, 10.0.

[5]These estimates are based on calculations using Connell's (1986) approximation to the LLNL one-dimensional atmospheric model.

[6]These conclusions are based on projected time paths of ozone concentration in Wuebbles (1983), Stordal and Isaksen (1986), and calculations using Connell's (1986) approximation to the LLNL model.

[7]For each probability of a bad information outcome (represented by $f_1^-(x)$), there is an optimal level of current-period regulations (including the possibility of no regulations). Since the calculations required to determine this optimal level require extensive iteration, critical probabilities are calculated for only a few representative levels of current-period regulations.

[8]Setting the surcharges proportional to relative ozone-depletion efficiencies is analogous to using taxes and subsidies to equalize the marginal private and social costs of an activity. The present value of the costs of limiting cumulative emissions over the planning period is minimized by increasing the surcharges at the discount rate, the optimal solution for pricing exhaustible resources (Fisher, 1981).

[9]The resource cost is the reduction in economic surplus due to the surcharge: the area bounded by the demand curve, the unregulated price, and the quantity demanded under the surcharge. See Camm et al. (1986) for further discussion.

[10]The choice of regulatory strategy can dramatically affect the distribution of consequences among consumers and firms, however. See Palmer and Quinn (February 1981, July 1981) for discussion of distributional effects and the relative advantages of various types of regulations.

[11]There are several differences between the demand curves employed in this simulation and those developed by Camm et al. (1986): (1) The simulation assumes that non-U.S. CFC-11 and 12 use as an aerosol propellant would begin to decline after their prices doubled (to $1.02/lb. and $1.32/lb.), and that use would thereafter decline linearly reaching only 5 percent of initial use if the price rose $5 per pound. (2) Camm et al. do not make any estimate of the reduction in demand for Halon-1211 and Halon-1301 at elevated prices. The simulation assumes their demand curves have constant elasticity equal to -0.32. The halons currently cost about $2/lb. Thus a $1/lb. increase would reduce simulated demand by 12 percent; a $5/lb. increase would reduce it 33 percent. (3) Camm et al. did not explicitly report technological options for reducing methyl chloroform emissions, but these are similar to those for CFC-113. The simulation uses a curve for methyl chloroform that is based on that for CFC-113 but adjusted for the difference in the price of the two chemicals. (4) Camm et al. also did not assess demand for the relatively minor uses of carbon tetrachloride other than use as a precursor to CFC-11 and 12. The simulations assume these other uses are

completely inelastic for price increases of less than $5/lb. (5) In some applications Camm et al. report two demand curves, depending on how widely options they identify as cost effective at current prices have been adopted. In these cases the simulated demand curves are half way between the two.

Chapter 10
Results and Sensitivity Analysis

[1]The weights used for calculating cumulative weighted emissions and the surcharges applied to each chemical, are proportional to their approximate estimated relative ozone-depletion efficiencies. The surcharge applied to methyl chloroform would be one-tenth the base surcharge listed in the text and the surcharges applied to the halons would be ten times the base surcharge.

[2]Current average U.S. prices per pound, based on United States International Trade Commission (1984), are: CFC-11, $0.51; CFC-12, $0.67; carbon tetrachloride, $0.16; and methyl chloroform, $0.30. The International Trade Commission does not publish data for CFC-113 or the halons but industry sources suggest the following average prices per pound: CFC-113, $0.89; Halon-1211, $1.95; and Halon-1301, $2.20.

[3]This value is calculated using the estimated average U.S. chemical prices noted above.

[4]The small, erratic variations are due to the varying curvature of the functions relating cumulative emissions to the required surcharge and associated resource costs (illustrated in Figs. 7 and 8) and approximations in simulating the demand curves. The algorithm approximates the demand curves with step functions that have steps at $0.05 intervals.

[5]These estimates correspond to the emission scenarios developed in Section A and the calculated ozone depletion shown in Fig. 4. Except for the point at 63.5 Mt, the emission paths corresponding to alternative regulations implicit in the figure differ from those used to calculate the projected ozone concentrations. However, these differences should not affect the projected ozone concentrations substantially.

[6]For the line labelled "1.0," the surcharge in year t satisfies the formula $s_{1.0}(t) = s^{1.0}$ x $(1.03)^{(t-1985)}$. For the other lines, labelled by the factor f, the surcharge in year t satisfies the formula $s_f(t) = f$ x s^f x $(1.03)^{(t-1985)}$ if $t < 1995$; $s_f(t) = s^f$ x $(1.03)^{(t-1985)}$ otherwise. Because the cumulative weighted emissions through the horizon are the same for each surcharge trajectory, the value of s^f depends on f and s^f is generally greater than $f s^{1.0}$.

[7]The modified demand function $q(p) = (1 + p/p_0)^\eta q_0(p)$, where $q(p)$ is the quantity demanded at surcharge p, $q_0(p)$ is the initial demand function, p_0 is the unregulated price of the chemical, and η is the incremental elasticity.

[8]These calculations assume an unregulated chemical price of $0.60/lb. which is approximately the price of CFC-11 and 12.

[9]Potential substitutes that are under U.S. regulatory scrutiny include the solvents perchloroethylene, trichloroethylene, and methylene chloride, which can substitute for CFC-113 and methyl chloroform in some applications. Methylene chloride can also be used in place of CFC-11 in manufacturing some flexible foams. Manzer (1990) discusses progress in developing new substances to replace PODs.

[10]The incremental elasticities in the other periods are: 1985-1990, 0.0; 1991-1995, -0.1; 1996-2000, -0.2; 2001-2005, -0.3; 2006-2010, -0.4; and 2011-2015, -0.5.

[11]These calculations assume an unregulated chemical price of $0.60/lb. as for the additional-elasticity case.

[12]The surcharges rise over time at the same discount rate as is used to measure the present value of resource costs. Thus the surcharges always minimize the present value of the resource cost of meeting any specified cumulative-emission limit.

Chapter 12
Valuing Health Risks

[1]The classic work on this approach is Schelling (1968). See also Acton (1976), Bailey (1980), Jones-Lee (1982), Fisher et al. (1986), and Viscusi (1983).

[2]For general discussion of valuation issues, see Rosen (1988), Smith (1990), and Smith and Desvousges (1988).

[3]The characterization of commodities as bundles of attributes is originally due to Lancaster (1966).

[4]See, for example, Atkinson and Halvorsen (1990), Bayless (1982), Blomquist (1979), Blomquist et al. (1988), Edmonds (1985), Freeman (1979a), Harrington et al. (1989), Harrison and Rubinfeld (1978), Hoehn, Berger and Blomquist (1984), Lucas (1977), Marin and Psacharopoulos (1982), Milon èt al. (1984), Ohta and Griliches (1976, 1986), Rosen (1981), Starr (1969), (Smith and Kaoru, 1987, 1990), Travis et al. (1987), Triplett (1976), Viscusi (1979, 1983), and Viscusi and Moore (1989).

[5]See Kahneman et al. (1982) for experimental evidence on humans' inability to accurately estimate risks and other quantities.

[6]See Clarke (1971), Groves (1973), Groves and Ledyard (1977), Hylland and Zeckhauser (1979), Leonard (1983), Mueller (1979), Satterthwaite (1975), Tideman and Tullock (1976), and Vickrey (1961).

[7]For a recent evaluation of the contingent-valuation method see Cummings et al. (1986a) or Mitchell and Carson (1989). Other references include Acton (1976), Bishop and Heberlein (1979), Brookshire and Coursey (1987), Brookshire and Crocker (1981), Brookshire et al. (1976, 1982), Desvousges et al. (1987), Gregory (1986), Gregory and McDaniels (1987), Hoehn and Randall (1987), Kealy et al. (1988), Knetsch and Sinden

(1984), Mitchell and Carson (1985, 1986), Reiling et al. (1990), Seller et al. (1985), Smith and Desvousges (1986, 1987), Smith et al. (1985), Violette and Chestnut (1983), and Viscusi et al. (1988).

[8]This conclusion depends on certain reasonable conditions, such as diminishing marginal returns to risk-reducing activities in each area. See Weinstein et al. (1980) for further discussion.

[9]Howard (1984) proposes using the value of a micromort—an incremental mortality risk of 10^{-6}, instead of the value of life. The use of micromorts may be preferred because it avoids the misleading impression that one would sell his or her life for the "value of life." However, the value of life has become the conventional measure for valuing risks so I adhere to this convention. The value of a micromort is simply 10^{-6} times the implicit value of life.

Chapter 13
Choosing Among Potentially Hazardous Foods

[1]DDT and dieldrin are banned in the United States, although DDT is still found as a constituent of dicophol. *Los Angeles Times*, August 13, 1985, Part I, p. 2; October 8, 1985, Part I, p. 3.

[2]California and Oregon are the only states that have laws regulating use of the label "organic." The term "natural" has no legal significance (MacFadyen, 1984).

[3]Hanemann (1984a) provides a rigorous development of econometric models of consumer demand involving simultaneous discrete and continuous choices, such as the choice of type and quantity of produce.

[4]When foods are strongly complementary it may be possible to define a composite commodity, the utility of which is additively separable from other foods.

[5]Oi (1973) derives similar conditions for consumer decisions concerning risky products in general. See also Goldberg (1974), who shows that if consumers are not perfectly informed, a ban on risky products may improve consumer welfare.

[6]If $\tau_i \equiv 0$ or, more generally, if $\tau_i(q_i) = \tau_i$ for all i such that $q_i > 0$.

[7]Risk ladders have been used recently to help survey respondents estimate the benefits of reduced environmental pollution. The risk ladder used is based on one developed by Mitchell and Carson (1986) to assess the benefits of reduced contamination of drinking water. Smith et al. (1985) were apparently the first to use risk ladders, in a study of aversion to living near a toxic-waste site.

[8]This difficulty may be explained by the "availability" heuristic. According to Tversky and Kahneman (1974), individuals' estimates of the likelihood of an event occurring may be influenced by how easily they can recall such an event. Slovic et al. (1978) argue that the availability heuristic helps to explain why few motorists wear seat belts: Serious automobile accidents are so infrequent that unbelted motorists nearly always survive each trip without incident.

Chapter 14
Willingness To Pay for Organically Grown Foods

[1]Two of the stores claim to use a fixed percentage markup for all products that is occasionally modified in accordance with prices at competing stores. The other stores declined to describe their markup policies.

[2]Several authors have studied the temporary depression of demand for specific foods following public announcement of their possible contamination. Van Ravenswaay and Hoehn (1990) studied the effects of new about Alar on apple sales, Foster and Just (1989) studied demand for Hawaiian milk contaminated by heptachlor, Johnson (1984) studied demand for potentially EDB-contaminated baked-goods mixes, Swartz and Strand (1981) studied regional demand for oysters when James River shellfish were possibly contaminated by kepone, and Shulstad and Stoevener (1978) studied the demand for pheasant hunting in Oregon when it was revealed that the birds were contaminated with mercury. Ippolito (1981) models a consumer's optimal reaction to new information that a food or other consumption good is more hazardous than previously believed. Schucker et al. (1983) analyze the effects of Saccharin warnings on diet soft drinks.

[3]Watermelon is a popular summer fruit often associated with Fourth of July picnics and barbeques, and hence demand may increase at that time. In July 1985 some California watermelons were found to be contaminated with aldicarb. Demand reportedly plummeted, but it was difficult to observe since the supply was sharply curtailed. All California-grown watermelons were removed from grocery shelves and restocking was allowed only from fields that were inspected and certified by the California Department of Food and Agriculture.

[4]Although there were 22 participants in the two organic-produce buyers' groups, two did not respond to the willingness-to-pay and food-risk questions. Consequently, the following tables show responses from only 20 organic-food purchasers.

[5]There is substantial experimental evidence that suggests individuals estimate an uncertain quantity by revising an estimate of some better known quantity (Kahneman et al., 1982). Typically, the adjustment away from the base is inadequate. Tversky and Kahneman (1974) characterize the resulting bias as "anchoring."

Chapter 15
Avoided Risk

[1]U.S. House of Representatives Agriculture Subcommittee on Department Operations, 1982 Staff Report, summarized in Mott (1985) and U.S. General Accounting Office (1986b).

[2]The literature on assessing human health risks from toxic chemicals, especially possible carcinogens, is large. See for example Albert (1985), Ames et al. (1987), Bernstein et al. (1985), Bolten et al. (1983), California Department of Health Services (1982), Crouch (1983), Crouch and Wilson (1979, 1981, 1982), Crump and Howe (1984), DuMouchel and Harris (1983), Food Safety Council (1980), Freedman and Navidi (1987, 1988), Freedman and Zeisel (1987), Haseman (1984, 1985), Hattis and Kennedy (1986), McElveen and Eddy (1985), Park and Snee (1983), Reynolds et al. (1987), Russell and Gruber (1987), Smith (1978), Tennant et al. (1987), Thorslund et al. (1987), Travis and White (1988), U.S. Environmental Protection Agency (August 1987), U.S. Interagency Staff Group on Carcinogens (1986), Wilson and Crouch (1987), and Zeise et al. (1984).

[3]When the LD_{50}s vary by sex the average of the two estimates is used.

[4]RTECS information was compared with other sources including California Department of Food and Agriculture (1985), Food and Drug Administration (1981), Gold et al. (1984), and Moses (1983). The sources are reasonably but not completely consistent.

[5]Pedersen et al. (1983) apply similar reasoning. An option in constructing their risk index for occupations or industries is to impute that a carcinogen is also a neoplastigen or vice versa. They do not impute carcinogenicity from mutagenicity but, for some types of chemicals, the Ames test and others that rely on mutagenicity have proven to be good indicators of carcinogenicity (cf. Crouch and Wilson, 1979).

[6]Pedersen et al.'s index also accounts for skin and eye irritation and other toxic effects such as to the liver or nervous system. Instead of simply recognizing whether a substance has been shown to have a particular effect it compares the compound's potency to all other substances reported to have that effect.

[7]$\psi_i = 1/2 \ (100/LD_{50}) + 1/2$ (weighting factors for the Chronic risk index).

[8]Ames et al. (1987) describe the Human Exposure/Rodent Potency (HERP) index, which is similar in spirit. It is the ratio of human exposure to the rodent TD_{50}.

[9]Gold et al. (1984) estimate TD_{50}s for each species, sex, and route of administration tested. The TD_{50}s reported in Table 13 approximate the median of the values reported for each pesticide.

[10]The median value of the Concentration risk index, which measures the average concentration of all pesticide residues, is 0.25 ppm. The estimate of 0.1 based on the Chronic risk index is equivalent to assuming that about 40 percent of pesticides, weighted by their average concentrations, are carcinogenic.

[11]A better method for calculating the average human intake of potentially carcinogenic pesticides would be to calculate the average concentration over all foods, weighted by the average quantities consumed. My approximation to this method, using the median concentration, may overestimate average intake: Ames (1983), citing U.S. FDA (undated), reports the average U.S. daily dietary intake of all synthetic organic pesticides as 60 μg. Using the median of the Concentration risk index (0.255 ppm), I estimate the same quantity as 255 μg (= 0.255 ppm x 1 kg), a factor of four greater. Part of the difference may be attributed to the 1 kg of produce I use as an average daily consumption. The average

American probably consumes less (especially during the 1970s, the period to which the FDA estimate applies), but the estimate may be reasonable for the individuals who explicitly choose to purchase organic produce, many of whom eat little or no meat. In addition, the median concentration is based on only 27 types of produce and does not include canned or other processed items. Pesticide concentrations may be higher in these 27 than in other types of produce, and higher in fresh than processed foods.

[12]Crouch and Wilson also report estimates for EDB based on commonly tested species of mice and rats: These are both 6 kg-day/mg, an order of magnitude higher than the human estimate.

[13]As reported in Table 13, Gold et al. estimate a TD_{50} for EDB of about 1 mg/kg body weight-day. Using this value in place of the median TD_{50} in the first estimate produces an estimated human risk 50 times as large, 7×10^{-4}, about two-thirds as large as the estimate based on the Crouch and Wilson potency measure.

[14]The average daily dose is 10^{-4} g (= 0.1 ppm x 1 kg of produce consumed). Again assuming that risk is proportional to dose, if a dose of 10^{-7} grams of aflatoxin is associated with a risk of 3×10^{-7}, a dose of 10^{-4} g should be associated with a risk of 3×10^{-4}.

[15]The rates are 2,040 for males, 1,650 for females (U.S. Department of Commerce, 1985).

Chapter 16
Willingness To Pay for Risk Reductions

[1]Participants in the focus groups seemed to indicate that there are no substantial systematic differences in taste between types. When differences exist, the organic items often taste better but look worse and are less consistent in size, shape, and color.

[2]Assuming the model specification is accurate, the t-statistic allows rejection of the hypothesis that the premium for organic grapefruit is negative, but not the hypothesis that none of the organic premiums are negative.

[3]The other foods with estimated premiums less than zero are lemons, cauliflower, and avocados. Since conventional avocados do not have detectable residues, one does not reduce his risk by purchasing organic avocados but does save money.

[4]Using the extreme values of a set of independent estimates for a set of parameters to estimate the values of the largest and smallest parameters will generally yield estimates that are biased towards the extremes. Assume one wishes to estimate the largest of a set of parameters $\{x_i\}$. For each x_i one has an estimate $y_i = x_i + \varepsilon_i$, where the ε_i are independent random variables with mean zero and nonzero variance. Even though each y_i is an unbiased estimate of the corresponding x_i, the largest y_i will overestimate the largest x_i, on average. Heuristically, one may select the wrong y_i. The largest y_i may be the largest not because it estimates the largest x_i but rather because ε_i happens to be large.

[5]It is convenient for calculation and exposition to employ this fiction of a once and for all choice. Consumers should switch between organically and conventionally grown produce as changes in prices and avoided risk shift the premium/avoided-risk ratios across their critical WTP value λ^*. Under the maintained assumption that risk is proportional to lifetime dose, the results of the two calculations are identical.

[6]If one measures the incremental risk reduction as an annual rather than a lifetime concept, it might be necessary to discount the future benefits as well as the costs (Keeler and Cretin, 1983).

[7]More accurately, the value of avoiding cancer. However, since the animal experiments used to estimate the TD_{50} generally do not distinguish between fatal and nonfatal cancers, the estimated risk of cancer is the same as that of fatal cancer.

[8]The implicit value of life is calculated by dividing the reported incremental WTP for the type of produce perceived as safer by the absolute value of the reported estimated risk difference. Thus, the conventional-food buyers categorized as having a negative WTP in Table 11 have a positive implicit value of life. This method of calculation differs from the method used for the revealed-preference estimates. There, the present value of incremental lifetime cost was divided by the avoided lifetime risk. Here, the annual incremental cost is divided by the annual avoided risk. The two methods are equivalent if the discount rate used in the revealed-preference estimates is zero.

[9]A sixth respondent also reported a WTP of zero, but estimated no difference in risk between the types of produce. The calculated implicit value of life, 0/0, is indeterminate and not included in the analysis.

[10]The focus groups were conducted in mid April, 1986. Under a law that became effective January 1 of that year, California motorists are required to wear seat belts. Participants were asked how frequently they had worn seat belts prior to the law. Since the law had been effective only three and one-half months at the time of the survey, recall problems may not be significant.

[11]Buss and Craik (1983) report that at least two distinct "world views" (constellations of attitudes and beliefs about world issues) have significant numbers of adherents. Adherents to alternative world views have differing perceptions of the severity of various technological and social risks. Douglas and Wildavsky (1982) analyze the manner in which social groups select the risks with which to be concerned.

Chapter 17
Conclusions

[1]The supermarkets in turn rely on government: The chief produce buyer for a major chain claims that he gives no attention at all to possible pesticide levels in choosing produce but relies on the government to ensure food safety (private communication, name withheld).

Chapter 18
Applying the Tools to Potential Climate Change

[1]For general background and discussion of the causes, characteristics, and consequences of climate change see Dickinson and Cicerone (1986), Clark (1988), National Research Council (1983), Pastor and Post (1988), Mitchell (1989), Oppenheimer and Boyle (1990), Penner (1990), Ramanathan et al. (1985), Titus (1986a, 1986b, 1986c, 1986d, 1988), Ramanathan (1988), Schneider (1989a, 1989b), Adams et al. (1990), Moomaw (1990), Singer (1989), Solow and Broadus (1990), Weiner (1990), and White (1990).

[2]Although widely used, the greenhouse analogy is somewhat misleading in that much of the effectiveness of a conventional greenhouse results from the physical containment of warmed air that would otherwise rise and be replaced by ambient air.

[3]Schlesinger (1984) and Schlesinger and Mitchell (1987) provide good reviews of climate modeling. See also Hansen et al. (1983, 1988) and Wilson and Mitchell (1987a, 1987b). Somerville (1987) cautions about the reliability of such models.

[4]Stavins and Jaffe (1990) estimate that 30 percent of forested wetland depletion in the lower Mississippi valley would not have occurred in the absence of federal flood-control projects.

BIBLIOGRAPHY

Acton, Jan Paul, "Measuring the Monetary Value of Lifesaving Programs," *Law and Contemporary Problems* 40, 1976.

Adams, R.M., S.A. Hamilton, and B.A. McCarl, "The Benefits of Pollution Control: The Case of Ozone and U.S. Agriculture," *American Journal of Agricultural Economics* 68: 886-893, 1986.

Adams, Richard M., Cynthia Rosenzweig, Robert M. Peart, Joe T. Ritchie, Bruce A. McCarl, J. David Glyer, R. Bruce Curry, James W. Jones, Kenneth J. Boote, and L. Hartwell Allen, Jr., "Global Climate Change and U.S. Agriculture," *Nature* 345: 219-224, 1990.

Adams, Richard M., Thomas D. Crocker, and Richard W. Katz, "Assessing the Adequacy of Natural Science Information: A Bayesian Approach," *Review of Economics and Statistics* 66: 568-575, 1984.

Adler, Robert S., and R. David Pittle, "Cajolery or Command: Are Education Campaigns an Adequate Substitute for Regulations?" *Yale Journal on Regulation* 1, 1984.

Adrian, John, and Raymond Daniel, "Impact of Socioeconomic Factors on Consumption of Selected Food Nutrients in the United States," *American Journal of Agricultural Economics* 57, February 1976.

Akerlof, George, and William T. Dickens, "The Economic Consequences of Cognitive Dissonance," *American Economic Review* 72, June 1982.

Albert, Roy E., "The Practical Importance of Antitumor Bioassay Responses in Carcinogenic Risk Assessment," *Risk Analysis* 5 (2), June 1985.

Alexander, Arthur, and Bridger M. Mitchell, *Measuring Technological Change of Heterogeneous Products*, R-3107-NSF, The RAND Corporation, Santa Monica, 1984.

Ames, Bruce N., "Dietary Carcinogens and Anticarcinogens," *Science* 221, 1256-1264, September 23, 1983.

Ames, Bruce N., Renae Magaw, and Lois Swirsky Gold, "Ranking Possible Carcinogenic Hazards," *Science* 236: 271-280, April 17, 1987.

Andelman, Julian B., and Dwight W. Underhill, eds., *Health Effects From Hazardous Waste Sites*, Lewis Publishers, Chelsea, Michigan, 1987.

Andersen, M.E., H.J. Clewell III, M.L. Gargas, F.A. Smith, and R.H. Reitz, "Physiologically Based Pharmacokinetics and the Risk Assessment Process for Methylene Chloride," *Toxicology and Applied Pharmacology* 87, 185-205, 1987.

Aneja, Viney P., "Natural Sulfur Emissions into the Atmosphere," *Journal of the Air and Waste Management Association* 40: 469-476, 1990.

Angelo, M.J., and A.B. Pritchard, "Simulations of Methylene Chloride Pharmacokinetics Using a Physiologically Based Model," *Regulatory and Toxicological Pharmacology* 4: 320-339, 1984.

Arrow, Kenneth J., "Risk Perception in Psychology and Economics," *Economic Inquiry* 20, 1982.

Arrow, Kenneth J., and Robert C. Lind, "Uncertainty and the Evaluation of Public Investment Decisions," *American Economic Review* 60, 364-378, 1970.

Arthey, V.D., *Quality of Horticultural Products*, Butterworth and Co., Ltd., London, 1975.

Arthur, W.B., "The Economics of Risks to Life," *American Economic Review* 71, 1981.

Asch, Peter, *Consumer Safety Regulation: Putting a Price on Life and Limb*, Oxford University Press, New York, 1988.

Assembly Office of Research, State of California, *The Leaching Fields: A Nonpoint Threat to Groundwater*, Sacramento, 1985.

Atkinson, Scott E., and Robert Halvorsen, "The Valuation of Risks to Life: Evidence from the Market for Automobiles," *Review of Economics and Statistics* 72: 133-136, 1990.

Ausubel, J.H., W.D. Nordhaus, "A Review of Estimates of Future Carbon Dioxide Emissions," in National Research Council, *Changing Climate: Report of the Carbon Dioxide Assessment Committee*, National Academy Press, 1983.

Ausubel, Jesse H., and Hedy E. Sladovich, eds., *Technology and Environment*, National Academy Press, Washington D.C., 1989.

Bailey, Martin J., *Reducing Risks to Life: Measurement of the Benefits*, American Enterprise Institute for Public Policy Research, Washington, D.C., 1980.

Baird, Brian N.R., "Tolerance for Environmental Health Risks: The Influence of Knowledge, Benefits, Voluntariness, and Environmental Attitudes," *Risk Analysis* 6: 425-435, 1986.

Barnett, T.P., "Beware Greenhouse Confusion," *Nature* 343: 696-697, 1990.

Barnola, J.M., D. Raynaud, Y.S. Korotkevich, and C. Lorius, "Vostok Ice Core Provides 160,000-Year Record of Atmospheric CO_2," *Nature* 329: 408-414, 1987.

Barr, A., and E.A. Feigenbaum, *The Handbook of Artificial Intelligence*, Vol. II, William Kaufman, Los Altos, California, 1982.

Barrett, C.R., and Prasanta K. Pattanaik, "Aggregation of Probability Judgements," *Econometrica* 55: 1237-1241, 1987.

Bates, J.A.R., and S. Gorbach, "Recommended Approach to the Appraisal of Risks to Consumers from Pesticide Residues in Crops and Food Commodities," *Pure and Applied Chemistry* 59: 611-624, 1987.

Bayless, Mark, "Measuring the Benefits of Air Quality Improvements: A Hedonic Salary Approach," *Journal of Environmental Economics and Management* 9, 1982.

Berger, Mark C., Glenn C. Blomquist, Don Kenkel, and George S. Trolley, *Valuing Changes in Health Risks: A Comparison of Alternative Measures*, University of Kentucky Working Paper in Economics, No. E-81-85, June 1985.

Bernstein, Leslie, Lois S. Gold, Bruce N. Ames, Malcolm C. Pike, and David G. Hoel, "Some Tautologous Aspects of the Comparison of Carcinogenic Potency in Rats and Mice," *Fundamental and Applied Toxicology* 5, 1985.

Berry, Donald A., and Bert Fristedt, *Bandit Problems: Sequential Allocation of Experiments*, Chapman and Hall, London, 1985.

Bettman, James R., and Mita Sujan, "Effects of Framing on Evaluation of Comparable and Noncomparable Alternatives by Expert and Novice Consumers," *Journal of Consumer Research* 14, 141-154, September 1987.

Bidwell, Robin, Frans Evers, Paul DeJongh, and Lawrence Susskind, "Public Perceptions and Scientific Uncertainty: The Management of Risky Decisions" *Environmental Impact Assessment Review* 7: 5-22, 1987.

Bikson, Tora Kay, and Jacqueline D. Goodchilds, *Product Decision Processes Among Older Adults*, R-2361-NSF, The RAND Corporation, Santa Monica, 1978.

Bishop, Richard C., and Thomas A. Heberlein, "Measuring Values of Extramarket Goods: Are Indirect Measures Biased?" *American Journal of Agricultural Economics* 61 (5), December 1979.

Blake, Donald R., and F. Sherwood Rowland, "Continuing Worldwide Increase in Tropospheric Methane, 1978 to 1987" *Science* 239: 1129-1131, 1988.

Blanciforti, L., R. Green, and S. Lane, "Income Expenditures for Relatively More versus Relatively Less Nutritious Foods over the Life Cycle," *American Journal of Agricultural Economics* 63 (2), 1981.

Blomquist, Glen, "The Value of Human Life: An Empirical Perspective," *Economic Inquiry* 19, 1981.

Blomquist, Glen, "Value of Life Saving: Implications of Consumption Activity," *Journal of Political Economy* 87, 1979.

Blomquist, Glenn C., Mark C. Berger, and John P. Hoehn, "New Estimates of Quality of Life in Urban Areas," *American Economic Review* 78: 89-107, 1988.

Blumthaler, Mario, and Walter Ambach, "Indication of Increasing Solar Ultraviolet-B Radiation Flux in Alpine Regions," *Science* 248: 206-208, 1990.

Bolten, J.G., P.F. Morrison, and K.A. Solomon, *Risk-Cost Assessment Methodology for Toxic Pollutants from Fossil Fuel Power Plants*, R-2993-EPRI, The RAND Corporation, Santa Monica, 1983.

Bolten, J.G., P.F. Morrison, K. Solomon, and K. Wolf, *Alternative Models for Risk Assessment of Toxic Emissions*, N-2261-EPRI, The RAND Corporation, Santa Monica, 1985.

Bowman, Kenneth P., "Global Trends in Total Ozone," *Science* 239, 48-50, 1 January 1988.

Brasseur, Guy, and Matthe W. Hitchman, "Stratospheric Response to Trace Gas Perturbations: Changes in Ozone and Temperature Distributions," *Science* 240: 634-637, 1988.

Brewer, Peter G., "Carbon Dioxide and the Oceans," in National Research Council, *Changing Climate: Report of the Carbon Dioxide Assessment Committee*, National Academy Press, 1983.

Breyer, Stephen, *Regulation and Its Reform*, Harvard University Press, Cambridge, 1982.

Broecker, Wallace S., "Unpleasant Surprises in the Greenhouse?" *Nature* 328: 123-126, 1987.

Broecker, Wallace S., Dorothy M. Peteet, David Rind, "Does the Ocean-Atmosphere System have More Than One Stable Mode of Operation?" *Nature* 315: 21-26, 1985.

Brookshire, David S., and Don L. Coursey, "Measuring the Value of a Public Good: An Empirical Comparison of Elicitation Procedures," *American Economic Review* 77: 554-566, 1987.

Brookshire, David S., and Thomas D. Crocker, "The Advantages of Contingent Valuation Methods for Benefit-Cost Analysis," *Public Choice* 36, 1981.

Brookshire, David S., Berry C. Ives, and William D. Schulze, "The Valuation of Aesthetic Preferences," *Journal of Environmental Economics and Management* 3, 1976.

Brookshire, David S., Mark A. Thayer, William D. Schulze, and Ralph C. d'Arge, "Valuing Public Goods: A Comparison of Survey and Hedonic Approaches," *American Economic Review* 72: 165-177, 1982.

Broome, John "The Economic Value of Life," *Economica* 52: 281-294, 1985.

Brown, James N., and Harvey S. Rosen, "On the Estimation of Structural Hedonic Price Models," *Econometrica* 50 (3), May 1982.

Brown, Peter G., "Policy Analysis, Welfare Economics, and the Greenhouse Effect," *Journal of Policy Analysis and Management* 7: 471-475, 1988.

Brown, Phil, "Popular Epidemiology: Community Response to Toxic Waste-Induced Disease in Woburn, Massachusetts," *Science, Technology, and Human Values* 12: 78-85, Summer/Fall 1987.

Bruce, Neil, and Richard G. Harris, "Cost-Benefit Criteria and the Compensation Principle in Evaluating Small Projects," *Journal of Political Economy* 90: 755-776, 1982.

Burkitt, Denis P., "Dietary Fiber as a Protection against Disease," in E.F. Patrice Jellife and Derrick B. Jellife, eds., *Adverse Effects of Foods*, Plenum Press, New York, 1982.

Burros, Marian, "Pure Food: The Status Symbol of the Decade; Who Buys It? The Affluent and the Aware," *New York Times*, April 2, 1986.

Busby, William F., and Gerald N. Wogan, "Aflatoxins," pp. 945-1134 in Charles E. Searle, ed., *Chemical Carcinogens*, 2d ed., American Chemical Society Monograph 182, Washington, D.C., 1984.

Buss, David M., and Kenneth H. Craik, "Contemporary World Views: Personal and Policy Applications," *Journal of Applied Social Psychology* 13: 259-280, 1983.

Butler, W.H., and J.M. Barnes, "Carcinogenic Action of Groundnut Meal Containing Aflatoxin in Rats," *Food and Cosmetics Toxicology* 6, 1968.

Butler, W.H., M. Greenblatt, and W. Lijinsky, "Carcinogenesis in Rats by Aflatoxins B_1, G_1, and B_2," *Cancer Research* 29, December 1969.

Cairns, John, "The Origin of Human Cancers," *Nature* 289, 353-357, January 29, 1981.

Cairns, John, "The Treatment of Diseases and the War against Cancer," *Scientific American* 253 (5), November 1985.

California Department of Food and Agriculture, *Birth Defects Prevention Act of 1984 (SB-950): Data Base Profile of Active Ingredients*, Sacramento, December 31, 1985.

California Department of Health Services, Epidemiological Studies Section, *Health Effects of Ethylene Dibromide*, Berkeley, 1985.

California Department of Health Services, *Carcinogen Identification Policy: A Statement of Science as a Basis of Policy, Section 2: Methods for Estimating Cancer Risks from Exposures to Carcinogens*, Sacramento, 1982.

Call, Gregory D., "Arsenic, ASARCO, and EPA: Cost-Benefit Analysis, Polluter Participation, and Polluter Games in the Regulation of Hazardous Air Pollutants," *Ecology Law Quarterly* 12: 567-617, 1985.

Camerer, Colin F., "Do Biases in Probability Judgment Matter in Markets? Experimental Evidence," *American Economic Review* 77: 981-997, December 1987.

Camm, Frank, and James K. Hammitt, *An Analytic Method for Constructing Scenarios from a Subjective Joint Probability Distribution*, N-2442-EPA, The RAND Corporation, Santa Monica, May 1986.

Camm, Frank, and Daniel F. Kohler, *Analyzing the Transitory Costs of Regulation with an Application to Toxic Chemicals*, N-2586-EPA, The RAND Corporation, Santa Monica, June 1987.

Camm, Frank, Timothy H. Quinn, Anil Bamezai, James K. Hammitt, Michael Meltzer, William E. Mooz, and Kathleen A. Wolf, *Social Cost of Technical Control Options to Reduce the Use of Potential Ozone Depleters in the United States: An Update*, N-2440-EPA, The RAND Corporation, Santa Monica, May 1986.

Cave, Jonathan A.K., *Age, Time, and the Measurement of Mortality Benefits*, R-3557-EPA, The RAND Corporation, Santa Monica, 1988.

Cess, R.D., G.L. Potter, J.P. Blanchet, G.J. Boer, S.J. Ghan, J.T. Kiehl, H. Le Treut, X.-Z. Liang, J.F.B. Mitchell, J.-J. Morcrette, D.A. Randall, M.R. Riches, E. Roeckner, U. Schlese, A. Slingo, K.E. Taylor, W.M. Washington, R.T. Wetherald, I. Yagai, "Interpretation of Cloud-Climate Feedback as Produced by 14 Atmospheric General Circulation Models," *Nature* 245: 513-516, 1989.

Chappellaz, J., J.M. Barnola, D. Raynaud, Y.S. Korotkevich, and C. Lorius, "Ice-Core Record of Atmospheric Methane over the Past 160,000 Years," *Nature* 345: 127-131, 1990.

Charlson, R.J., J.E. Lovelock, M.O. Andreae, and S.J. Warren, "Oceanic Phytoplankton, Atmospheric Sulfur, Cloud Albedo, and Climate," *Nature* 326: 655-661, 1987.

Chemical Manufacturers Association, *Production, Sales, and Calculated Release of CFC-11 and CFC-12 through 1985*, Washington, D.C., 1986.

Cicerone, Ralph J., "Changes in Stratospheric Ozone," *Science* 237: 35-42, 1987.

Cicerone, Ralph J., Richard S. Stolarski, and Stacy Walters, "Stratospheric Ozone Destruction by Man-Made Chlorofluoromethanes," *Science* 185: 1165-1167, 1974.

Clark, James S., "Effect of Climate Change on Fire Regimes in Northwestern Minnesota," *Nature* 334: 233-235, 1988.

Clarke, Edward H., "Multipart Pricing of Public Goods," *Public Choice* 11, 17-33, Fall 1971.

Clawson, Marion, and Jack Knetsch, *Economics of Outdoor Recreation*, Johns Hopkins University Press, Baltimore, 1966.

Coffin, D.E., and W.P. McKinley, "Sources of Pesticide Residues," in Horace D. Graham, ed., *The Safety of Foods*, 2d ed., AVI Publishing Co., Westport, Connecticut, 1980.

Connell, Peter S., *A Parameterized Numerical Fit to Total Column Ozone Changes Calculated by the LLNL 1-D Model of the Troposphere and Stratosphere*, Lawrence Livermore National Laboratory, Informal Report UCID-20762, May 1986.

Connell, Peter S., and Donald J. Wuebbles, *Ozone Perturbations in the LLNL One-Dimensional Model—Calculated Effects of Projected Trends in CFCs, CH_4, CO_2, N_2O, and Halons over 90 Years*, UCRL-95548, Lawrence Livermore National Laboratory, Livermore, 1986.

Conrad, Jon M., "Quasi-Option Value and the Expected Value of Information," *Quarterly Journal of Economics* 92: 813-820, June 1980.

Cook, Philip J., and Daniel A. Graham, "The Demand for Insurance and Protection: The Case of Irreplaceable Assets," *Quarterly Journal of Economics* 91, 143-154, February 1977.

Cosslett, Stephen R., "Distribution-Free Maximum Likelihood Estimator of the Binary Choice Model," *Econometrica* 51 (3), May 1983.

Coursey, Don L., John L. Hovis, and William D. Schulze, "The Disparity between Willingness To Accept and Willingness To Pay Measures of Value," *Quarterly Journal of Economics* 102: 679-695, 1987.

Covello, Vincent T., Detlof von Winterfeldt, and Paul Slovic, "Risk Communication: A Review of the Literature," *Risk Abstracts* 3: 171-181, October 1986.

Covello, Vincent T., W. Gary Flamm, Joseph V. Rodricks, and Robert G. Tardiff, *The Analysis of Actual Versus Perceived Risks*, Plenum Press, New York, 1983.

Cowling, Keith, and A.J. Rayner, "Price, Quality and Market Share," *Journal of Political Economy* 78, 1970.

Crawford, Mark, "Landmark Ozone Treaty Negotiated," *Science* 237: 1557, 1987.

Crosson, Pierre R., "Climate Change: Problems of Limits and Policy Responses," in N.J. Rosenberg, W.E. Easterling III, P.R. Crosson, and J. Darmstadter, eds., *Greenhouse Warming: Abatement and Adaptation*, Resources for the Future, Washington, D.C., 1989.

Crouch, Edmund A.C., "Uncertainties in Interspecies Extrapolations of Carcinogenicity," *Environmental Health Perspectives* 50, 1983.

Crouch, Edmund A.C., and Richard Wilson, "Interspecies Comparison of Carcinogenic Potency," *Journal of Toxicology and Environmental Health* 5, 1979.

Crouch, Edmund A.C., and Richard Wilson, "Regulation of Carcinogens," *Risk Analysis* 1 (1), 1981.

Crouch, Edmund A.C., and Richard Wilson, *Risk/Benefit Analysis*, Ballinger Publishing Co., Cambridge, Massachusetts, 1982.

Crump, Kenny S., and Richard B. Howe, "The Multistage Model with a Time-Dependent Dose Pattern: Applications to Carcinogenic Risk Assessment," *Risk Analysis* 4: 163-176, 1984.

Cummings, R., W. Schulze, S. Gerking, and D. Brookshire, "Measuring the Elasticity of Substitution of Wages for Municipal Infrastructure: A Comparison of the Survey and Wage Hedonic Approaches," *Journal of Environmental Economics and Management* 13, 1986b.

Cummings, R.G., D.S. Brookshire, and W.D. Schulze, *Valuing Environmental Goods: An Assessment of the Contingent Valuation Method*, Rowman and Allanheld, Totowa, New Jersey, 1986a.

Cyert, R.M., M.H. DeGroot, and C.A. Holt, "Sequential Investment Decisions with Bayesian Learning," *Management Science* 24 (7), 1978.

Darby, William P., "An Example of Decision-Making on Environmental Carcinogens: The Delaney Clause," *Journal of Environmental Systems* 9: 109-121, 1979-1980.

Deaton, Angus, and John Muellbauer, *Economics and Consumer Behavior*, Cambridge University Press, New York, 1980.

DeGroot, Morris H., *Optimal Statistical Decisions*, McGraw-Hill, New York, 1970.

DeMare, Jacques, "Optimal Prediction of Catastrophes with Applications to Gaussian Processes," *The Annals of Probability* 8: 841-850, 1980.

DeMore, W.B., ed., *Chemical Kinetics and Photochemical Data for Use in Stratospheric Modeling*, Jet Propulsion Laboratory Publication 85-37, Pasadena, California, 1985.

Desvousges, William H., *Radon Focus Groups: A Summary*, Research Triangle Institute, Research Triangle Park, North Carolina, 1986.

Desvousges, William H., and V. Kerry Smith, "Focus Groups and Risk Communication: The 'Science' of Listening to Data," *Risk Analysis* 8: 479-484, 1988.

Desvousges, William H., V. Kerry Smith, and Ann Fisher, "Option Price Estimates for Water Quality Improvements: A Contingent Valuation Study for the Monongahela River," *Journal of Environmental Economics and Management* 14: 248-267, 1987.

Detwiler, R.P., and Charles A.S. Hall, "Tropical Forests and the Global Carbon Cycle," *Science* 239: 42-47, 1988.

Dewees, Donald N., and Thomas A. Wilson, "Cold Houses and Warm Climates Revisited: On Keeping Warm in Chicago, or Paradox Lost," *Journal of Political Economy* 98: 656-663, 1990.

deZafra, R.L., M. Jaramillo, A. Parrish, P. Solomon, B. Connor, and J. Barrett, "High Concentrations of Chlorine Monoxide at Low Altitudes in the Antarctic Spring Stratosphere: Diurnal Variation," *Nature* 328: 408-411, 1987.

Dichter, C.R., "Risk Estimates of Liver Cancer due to Aflatoxin Exposure from Peanuts and Peanut Products," *Food and Cosmetics Toxicology* 22: 431-437, 1984.

Dickie, Mark, Ann Fisher, and Shelby Gerking, "Market Transactions and Hypothetical Demand Data: A Comparative Study," unpublished, University of Wyoming, November 1985.

Dickinson, Robert E., and Ralph J. Cicerone, "Future Global Warming from Atmospheric Trace Gases," *Nature* 319: 109-115, 1986.

Doll, Richard, and Richard Peto, "The Causes of Cancer: Quantitative Estimates of Avoidable Risks of Cancer in the United States Today," *Journal of the National Cancer Institute* 66, 1981.

Dorfman, Robert, "Incidence of the Benefits and Costs of Environmental Programs," *American Economic Review* 67: 333-340, 1977.

Douglas, Mary, and Aaron Wildavsky, *Risk and Culture: An Essay on the Selection of Technical and Environmental Dangers*, University of California Press, Berkeley, 1982.

Doull, John, Curtis D. Klaassen, and Mary O. Amdur, *Casarett and Doull's Toxicology: The Basic Science of Poisons*, 2d. ed., Macmillan Publishing Co., Inc., New York, 1980.

Dowell, Richard, "Risk Preference and the Work-Leisure Trade-Off," *Economic Inquiry* 23, October 1985.

Downing, D.J., R.H. Gardner, F.O. Hoffman, "An Examination of Response-Surface Methodologies for Uncertainty Analysis in Assessment Models," *Technometrics* 27: 151-163, 1985.

Draper, David, James S. Hodges, Edward E. Leamer, Carl N. Morris, and Donald B. Rubin, *A Research Agenda for Assessment and Propagation of Model Uncertainty*, N-2683-RC, The RAND Corporation, Santa Monica, November 1987.

Duan, Naihua, *Application of the Microenvironment Approach to Assess Human Exposure to Carbon Monoxide*, R-3222-EPA, The RAND Corporation, Santa Monica, January 1985.

Duggan, Reo E., Paul E. Corneliussen, Mary B. Duggan, Bernadette M. McMahon, and Robert J. Martin, *Residue Monitoring Data: Pesticide Residue Levels in Foods in the United States from July 1, 1969 to June 30, 1976*, Food and Drug Administration and the Association of Official Analytical Chemists, Washington D.C., 1983.

DuMouchel, William H., and Jeffrey E. Harris, "Bayes Methods for Combining the Results of Cancer Studies in Humans and Other Species," with comments, *Journal of the American Statistical Association* 78: 293-315, 1983.

Eckstein, Zvi, Martin Eichenbaum, and Dan Peled, "Uncertain Lifetimes and the Welfare Enhancing Properties of Annuity Markets and Social Security," *Journal of Public Economics* 26, 1985.

Edgerton, Sylvia A., Kirk R. Smith, Richard A. Carpenter, Toufiq A. Siddiqi, Steven G. Olive, Corazon Pe Benito Claudio, Vincent T. Covello, Donald J. Fingleton, Kwi-Gon Kim, and Bruce A. Wilcox, "Priority Topics in the Study of Environmental Risk in Developing Countries: Report on a Workshop Held at the East-West Center, August, 1988," *Risk Analysis* 10: 273-283, 1990.

Edmonds, J.A., J. Reilly, J.R. Trabalka, D.E. Reichle, *An Analysis of Possible Future Atmospheric Retention of Fossil Fuel CO_2*, Technical Report TR013, U.S. Department of Energy Carbon Dioxide Research Division, 1984.

Edmonds, Radcliffe G., "Some Evidence on the Intertemporal Stability of Hedonic Price Function," *Land Economics* 61 (4), November 1985.

Edwards, Steven F., and Glen D. Anderson, "Overlooked Biases in Contingent Valuation Surveys: Some Considerations," *Land Economics* 63: 168-178, 1987.

Ehrlich, I., and G.S. Becker, "Market Insurance, Self-Insurance, and Self-Protection," *Journal of Political Economy* 80: 623-648, 1972.

Elfner, Douglas, and K. Celeste Gaspari, "Estimating the Impact of an Unobservable Variable: The Impact of Lifestyle on Health Status," unpublished paper presented at the December 1984 American Economics Association meetings.

Epple, Dennis, "Hedonic Prices and Implicit Markets: Estimating Demand and Supply Functions for Differentiated Products," unpublished, Graduate School of Industrial Administration, Carnegie-Mellon University, April 1984.

Epstein, Samuel S., *The Politics of Cancer*, Anchor Press/Doubleday, Garden City, New York, 1979.

Farm Chemicals Handbook '85, Meister Publishing Co., Willoughby, Ohio, 1985.

Farman, J.C., B.G. Gardiner, and J.D. Shanklin, "Large Losses of Total Ozone in Antarctica Reveal Seasonal ClO_x/NO_x Interaction," *Nature* 315: 207-210, 16 May 1985.

Farmer, C.B., G.C. Toon, P.W. Schaper, J.-F. Blavier, and L.L. Lowes, "Stratospheric Trace Gases in the Spring 1986 Antarctic Atmosphere," *Nature* 329, 126-130, 10 September 1987.

Feenburg, Daniel, and Edwin S. Mills, *Measuring the Benefits of Water Pollution Abatement*, Academic Press, New York, 1980.

Fessenden-Raden, June, Janet M. Fitchen, and Jenifer S. Heath, "Providing Risk Information in Communities: Factors Influencing What Is Heard and Accepted," *Science, Technology, and Human Values* 12, 94-101, Summer/Fall 1987.

Finkel, Adam M., *Confronting Uncertainty in Risk Management: A Guide for Decision Makers*, Resources for the Future, Washington, D.C., 1990.

Fisher, Ann, Lauraine G. Chestnut, and Daniel M. Violette, "The Value of Reducing Risks of Death: A Note on New Evidence," *Journal of Policy Analysis and Management* 8: 88-100, 1989.

Fisher, Anthony C., *Resource and Environmental Economics*, Cambridge University Press, Cambridge, 1981.

Fisher, Donald A., Charles H. Hales, David L. Filkin, Malcolm K.W. Ko, N. Dak Sze, Peter S. Connell, Donald J. Wuebbles, Ivar S.A. Isaksen, and Frode Stordal, "Model Calculations of the Relative Effects of CFCs and their Replacements on Stratospheric Ozone," *Nature* 344: 508-512, 1990a.

Fisher, Donald A., Charles H. Hales, Wei-Chyung Wang, Malcolm K.W. Ko, and N. Dak Sze, "Model Calculations of the Relative Effects of CFCs and their Replacements on Global Warming," *Nature* 344: 513-516, 1990b.

Food Marketing Institute, *Trends: Consumer Attitudes and the Supermarket*, Washington, D.C., 1986.

Food Safety Council, "Quantitative Risk Assessment," *Food and Cosmetics Toxicology* 18: 711-734, 1980.

Foster, William, and Richard E. Just, "Measuring Welfare Effects of Product Contamination with Consumer Uncertainty," *Journal of Environmental Economics and Management* 17: 266-283, 1989.

Franco, Jacques, "An Analysis of the California Market for Organically Grown Produce," *American Journal of Alternative Agriculture* 4: 22-27, 1989.

Fraser, Clive D., "Optimal Compensation for Potential Fatality," *Journal of Public Economics* 23, 1984.

Frederick, John E., and Hilary E. Snell, "Ultraviolet Radiation Levels During the Antarctic Spring," *Science* 241: 438-440, 1988.

Freedman, David A., and Hans Zeisel, "From Mouse to Man: The Quantitative Assessment of Cancer Risks," *Statistical Science* (forthcoming with comments), Technical Report No. 79, Department of Statistics, University of California, Berkeley, July 1, 1987.

Freedman, David A., and W. Navidi, "On the Multistage Model for Carcinogenesis," Technical Report No. 97, Statistics Department, University of California, Berkeley, December 1, 1987.

Freedman, David A., and W. Navidi, "On the Risk of Lung Cancer for Ex-Smokers," Technical Report No. 135, Statistics Department, University of California, Berkeley, January 30, 1988.

Freeman, A. Myrick III, "Estimating the Benefits of Environmental Regulations," in Arthur D. Little, Inc., *Evaluation of State-of-the-Art in Benefit Assessment Methods for Public Policy Purposes*, December 1984.

Freeman, A. Myrick III, "Hedonic Prices, Property Values and Measuring Environmental Benefits: A Survey of the Issues," *Scandinavian Journal of Economics* 81 (2), 1979a.

Freeman, A. Myrick III, *The Benefits of Environmental Improvement: Theory and Practice*, Johns Hopkins University Press, Baltimore, 1979b.

Friedman, David, "Cold Houses in Warm Climates and Vice Versa: A Paradox of Rational Heating," *Journal of Political Economy* 95: 1089-1097, 1987.

Froines, John R., Cornelia A. Dellenbaugh, Sharon S. Seabrook, and David H. Wegman, *A Profile of Occupational Health Experience in Los Angeles County*, Southern Occupational Health Center, School of Public Health, University of California, Los Angeles, 1984.

Fuchs, Victor R., and Richard Zeckhauser, "Valuing Health—A 'Priceless' Commodity," *American Economic Association Papers and Proceedings* 77: 263-268, 1987.

Gallagher, David R., and V. Kerry Smith, "Measuring Values for Environmental Resources under Uncertainty," *Journal of Environmental Economics and Management* 12, 132-143, 1985.

Gamlen, P.H., B.C. Lane, P.M. Midgley, and J.M. Steed, "The Production and Release to the Atmosphere of CCl_3F and CCl_2F_2 (Chlorofluorocarbons CFC 11 and CFC 12)," *Atmospheric Environment*, 12: 1077-1085, 1986.

Gardner, Geral T., Adrian R. Tiemann, Leroy C. Gould, Donald R. DeLuca, Leonard W. Doob, and Jan A.J. Stolwijk, "Risk and Benefit Perceptions, Acceptability Judgments, and Self-Reported Actions Toward Nuclear Power," *Journal of Social Psychology* 116: 197-197, 1982.

Geller, Marvin A., "Solar Cycles and the Atmosphere," *Nature* 332: 584-585, 1988.

Gerking, Shelby, and Linda R. Stanley, "An Economic Analysis of Air Pollution and Health: The Case of St. Louis," *Review of Economics and Statistics* 68: 115-121, February 1986.

Gillick, Muriel R., "Common-Sense Models of Health and Disease," *The New England Journal of Medicine* 313 (11), September 12, 1985.

Gleick, James, *Chaos: Making a New Science*, Heinemann, London, 1988.

Gleick, Peter H., "Climate Change, Hydrology, and Water Resources," *Reviews of Geophysics* 27: 329-344, 1989.

Gold, Lois Swirsky, Charles B. Sawyer, Renae Magaw, Georganne M. Backman, Margarita de Veciana, Robert Levinson, N. Kim Hooper, William R. Havender, Leslie Bernstein, Richard Peto, Malcolm C. Pike, and Bruce N. Ames, "A Carcinogenic Potency Database of the Standardized Results of Animal Bioassays," *Environmental Health Perspectives* 58, 1984.

Goldbaum, Gary M., Patrick L. Remington, Kenneth E. Powell, Gary C. Hogelin, Eileen M. Gentry, The Behavioral Risk Factor Surveys Group, "Failure to Use Seat Belts in the United States," *Journal of the American Medical Association* 255: 2459-2462, 1986.

Goldberg, Victor P., "The Economics of Product Safety and Imperfect Information," *Bell Journal of Economics and Management Science* 5 (2), Autumn 1974.

Goldemberg, Jose, Thomas B. Johansson, Amulya K.N. Reddy, and Robert H. Williams, *Energy for a Sustainable World*, World Resources Institute, Washington, D.C., 1987.

Goodin, Robert E., "Discounting Discounting," *Journal of Public Policy* 2: 53-72, February 1982.

Graham, Daniel A., "Cost-Benefit Analysis under Uncertainty," *American Economic Review* 71: 715-725, September 1981.

Graham, John D., and Steven Garber, "Evaluating the Effects of Automobile Safety Regulation," *Journal of Policy Analysis and Management* 3: 206-224, 1984.

Grant, William B., "Global Stratospheric Ozone and UVB Radiation," (with response) *Science* 242: 1111-1112, 1988.

Gregory, Robin, "Interpreting Measures of Economic Loss: Evidence from Contingent Valuation and Experimental Studies," *Journal of Environmental Economics and Management* 13: 325-337, 1986.

Gregory, Robin, and Tim McDaniels, "Valuing Environmental Losses: What Promise Does the Right Measure Hold?" *Policy Sciences* 20, 11-26, 1987.

Grether, David M., and Charles R. Plott, "Economic Theory of Choice and the Preference Reversal Phenomenon," *American Economic Review* 69 (4), September 1979.

Grossman, Sanford J., Richard E. Kihlstrom, and Leonard J. Mirman, "A Bayesian Approach to the Production of Information and Learning by Doing," *Review of Economic Studies* 44: 533-547, 1977.

Groves, T., "Incentives in Teams," *Econometrica* 41, July 1973.

Groves, T., and J. Ledyard, "Optimal Allocation of Public Goods: A Solution to the 'Free Rider' Problem," *Econometrica* 45, May 1977.

Haaga, John, "Children's Seatbelt Usage: Evidence from the National Health Interview Survey," *American Journal of Public Health* 76: 1425-1427, 1986.

Hall, Darwin C., Brian P. Baker, Jacques Franco, and Desmond A. Jolly, "Organic Food and Sustainable Agriculture," *Contemporary Policy Issues* 7: 47-72, 1989.

Hambraeus, Leif, "Naturally Occurring Toxicants in Food," in E.F. Patrice Jellife and Derrick B. Jellife, eds., *Adverse Effects of Foods*, Plenum Press, New York, 1982.

Hamermesh, Daniel S., "Expectations, Life Expectancy, and Economic Behavior," *Quarterly Journal of Economics*, May 1985.

Hammitt, James K., "The Timing of Regulations to Prevent Stratospheric-Ozone Depletion," (oral presentation), United Nations Environment Programme Workshop on the Control of Chlorofluorocarbons, Leesburg, Virginia, September 1986.

Hammitt, James K., *Estimating Consumer Willingness to Pay to Reduce Food-Borne Risk*, R-3447-EPA, The RAND Corporation, Santa Monica, October 1986.

Hammitt, James K., *Timing Regulations to Prevent Stratospheric-Ozone Depletion*, R-3495-JMO/RC, The RAND Corporation, Santa Monica, April 1987.

Hammitt, James K., "Subjective-Probability-Based Scenarios for Uncertain Input Parameters: Stratospheric Ozone Depletion," *Risk Analysis* 10: 93-102, 1990; also published as N-3140-EPA/JMO/RC, The RAND Corporation, Santa Monica, April 1990a.

Hammitt, James K., "Risk Perceptions and Food Choice: An Exploratory Analysis of Organic- Versus Conventional-Produce Buyers," *Risk Analysis* 10: 367-374, 1990b.

Hammitt, James K., and Jonathan A.K. Cave, *Research Planning for Food Safety: A Value-of-Information Approach*, R-3946-ASPE/NCTR, The RAND Corporation, Santa Monica, 1990.

Hammitt, James K., Kathleen A. Wolf, Frank Camm, William E. Mooz, Timothy H. Quinn, and Anil Bamezai, *Product Uses and Market Trends for Potential Ozone-Depleting Substances, 1985-2000*, R-3386-EPA, The RAND Corporation, Santa Monica, May 1986.

Hammitt, James K., Frank Camm, Peter S. Connell, William E. Mooz, Kathleen A. Wolf, Donald J. Wuebbles, and Anil Bamezai, "Joint Emission Scenarios for Potential Stratospheric-Ozone Depleting Substances," *Nature* 330: 711-716, 24/31 December 1987; also published as R-3628-JMO/RC, The RAND Corporation, Santa Monica, March 1988.

Hammitt, James K., Jonathan A.K. Cave, Muhammad G. Mustafa, and R. Burciaga Valdez, *Research Planning for Food Safety: Preliminary Methodology and Applications*, N-2836-ASPE/NCTR/JMO, The RAND Corporation, Santa Monica, March 1989.

Hammond, Peter J., "Changing Tastes and Coherent Dynamic Choice," *Review of Economic Studies* 43, 159-173, 1976.

Hanemann, W. Michael, "Discrete/Continuous Models of Consumer Demand," *Econometrica* 52 (3), May 1984a.

Hanemann, W. Michael, "Willingness to Pay and Willingness to Accept: How Much Can They Differ?" Working Paper No. 328, Division of Agricultural Sciences, University of California, Berkeley, August 1984b.

Hansen, James, and Sergej Lebedeff, "Global Trends of Measured Surface Air Temperature," *Journal of Geophysical Research* 92 (D11): 13345-13372, 1987.

Hansen, J., I. Fung, A. Lacis, D. Rind, S. Lebedeff, R. Ruedy, G. Russell, and P. Stone, "Global Climate Changes as Forecast by Goddard Institute for Space Studies Three-Dimensional Model," *Journal of Geophysical Research* 93 (D8): 9341-9364, 1988.

Hansen, J., G. Russell, D. Rind, P. Stone, A. Lacis, S. Lebedeff, R. Ruedy, and L. Travis, "Efficient Three-Dimensional Global Models for Climate Studies: Models I and II," *Monthly Weather Review* 111: 609-662, 1983.

Harrington, Winston, and Paul R. Portney, "Valuing the Benefits of Health and Safety Regulation," *Journal of Urban Economics* 22, 101-112, 1987.

Harrington, Winston, Alan J. Krupnick, and Walter O. Spofford, Jr., "The Economic Losses of a Waterborne Disease Outbreak," *Journal of Urban Economics* 25: 116-137, 1989.

Harrison, David Jr., and Daniel L. Rubinfeld, "Hedonic Housing Prices and the Demand for Clean Air," *Journal of Environmental Economics and Management* 5, 1978.

Haseman, Joseph K., "Evaluating the Carcinogenic Potential of a Chemical that Appears to Both Increase and Decrease Tumor Incidences," *Risk Analysis* 5 (2), June 1985.

Haseman, Joseph K., "Statistical Issues in the Design, Analysis and Interpretation of Animal Carcinogenicity Studies, *Environmental Health Perspectives* 58, 1984.

Hattis, Dale, and David Kennedy, "Assessing Risks from Health Hazards: An Imperfect Science," *Technology Review* 89 (4), May/June, 1986.

Hayes, Wayland Jr., *Pesticides Studied in Man*, Williams and Wilkins Co., Baltimore, 1982.

Hearne, Shelly A., *Harvest of Unknowns: Pesticide Contamination in Imported Foods*, Natural Resources Defense Council, San Francisco, 1984.

Heath, D.F., "Non-Seasonal Changes in Total Column Ozone from Satellite Observations: 1970-1985," *Nature* 332: 219-227, 17 March 1988.

Henrion, Max, and Baruch Fischhoff, "Assessing Uncertainty in Physical Constants," *American Journal of Physics* 54: 791-798, 1986.

Hey, John D., "Are Optimal Search Rules Reasonable? And Vice Versa? (And Does It Matter Anyway?)," *Journal of Economic Behaviour and Organization* 2: 47-70, 1981.

Hilgartner, Stephen, and Dorothy Nelkin, "Communication Controversies over Dietary Risks," *Science, Technology, and Human Values* 12, 41-47, Summer/Fall 1987.

Hills, Alan J., Ralph J. Cicerone, Jack G. Calvert, and John W. Birks, "Kinetics of the BrO + ClO reaction and implications for Stratospheric Ozone," *Nature* 328: 405-408, 1987.

Hirshleifer, Jack, *Price Theory and Applications*, Prentice-Hall, Inc., Englewood Cliffs, N.J., 1976

Hirshleifer, Jack, and John G. Riley, "The Analytics of Uncertainty and Information—An Expository Survey," *Journal of Economic Literature* 17: 1375-1421, December 1979.

Hodges, James S., "Uncertainty, Policy Analysis and Statistics," *Statistical Science* 2: 259-291, 1987.

Hoehn, John P., and Alan Randall, "A Satisfactory Benefit Cost Indicator from Contingent Valuation," *Journal of Environmental Economics and Management* 14: 226-247, 1987.

Hoehn, John P., Mark C. Berger, and Glenn C. Blomquist, *A Multimarket Approach to Valuing Environmental Amenities*, Staff Paper No. 84-68, Department of Agricultural Economics, Michigan State University, December 1984.

Hogarth, Robin M., and Howard Kunreuther, "Ambiguity and Insurance Decisions," *American Economic Review* 75 (2), May 1985.

Holt, Charles A., "Preference Reversals and the Independence Axiom," *American Economic Review* 76 (3), June 1986.

Hough, Adrian M., and Roger G. Derwent, "Changes in the Global Concentration of Tropospheric Ozone due to Human Activities," *Nature* 344: 645-648, 1990.

Howard, Ronald A., "On Fates Comparable to Death," *Management Science* 30, 1984.

Howard, Ronald A., "Uncertainty About Probability," *Risk Analysis* 8: 91-98, March 1988.

Howe, Charles W., and K. William Easter, *Interbasin Transfers of Water: Economic Issues and Impacts*, Johns Hopkins Press, Baltimore, 1971.

Hylland, Aanund, and Richard Zeckhauser, "The Efficient Allocation of Individuals to Positions," *Journal of Political Economy* 87: 293-314, 1979.

Iman, R.L., J.C. Helton, "An Investigation of Uncertainty and Sensitivity Analysis Techniques for Computer Models," *Risk Analysis* 8: 71-90, 1988.

Ippolito, Pauline M., "Information and the Life Cycle Consumption of Hazardous Goods," *Economic Inquiry* 19, 1981.

Ippolito, Pauline M., and Richard A. Ippolito, "Measuring the Value of Life Saving from Consumer Reactions to New Information," *Journal of Public Economics* 25, 1984.

Isaksen, Ivar S.A., and Frode Stordal, "Ozone Perturbations by Enhanced Levels of CFCs, N_2O, and CH_4: A Two-Dimensional Diabatic Circulation Study Including Uncertainty Estimates," *Journal of Geophysical Research* 91 (D4): 5249-5263, 1986.

Jacoby, Jacob, George J. Szybillo, and Jacqueline Busato-Schach, "Information Acquisition Behavior in Brand Choice Situations," *Journal of Consumer Research* 3, 1977.

Jacoby, Jacob, Jerry C. Olson, and Rafael A. Haddock, "Price, Brand Name, and Product Composition Characteristics as Determinants of Perceived Quality," *Journal of Applied Psychology* 55, 1971.

Jacoby, Jacob, Robert W. Chestnut, and William Silberman, "Consumer Use and Comprehension of Nutrition Information," *Journal of Consumer Research* 4, 1977.

Jellife, E.F. Patrice, and Derrick B. Jellife, eds., *Adverse Effects of Foods*, Plenum Press, New York, 1982.

Jodha, N.S., "Potential Strategies for Adapting to Greenhouse Warming: Perspectives from the Developing World," in N.J. Rosenberg, W.E. Easterling III, P.R. Crosson, and J. Darmstadter, eds., *Greenhouse Warming: Abatement and Adaptation*, Resources for the Future, Washington, D.C., 1989.

Johnson, F. Reed, "Market Disruption, Regulated Health Risks, and the Media: The Case of EDB," unpublished, U.S. Environmental Protection Agency, Washington, D.C., December 1984.

Johnson, F. Reed, and Ralph A. Luken, "Radon Risk Information and Voluntary Protection: Evidence from a Natural Experiment," *Risk Analysis* 7: 97-107, 1987.

Jones, P.D., T.M.L. Wigley, and P.B. Wright, "Global Temperature Variations Between 1861 and 1984," *Nature* 322: 430-434, 1986.

Jones, P.D., T.M.L. Wigley, C.K. Folland, D.E. Parker, J.K. Angell, S. Lebedeff, and J.E. Hansen, "Evidence for Global Warming in the Past Decade," *Nature* 332: 790, 1988.

Jones-Lee, Michael W., ed., *The Value of Life and Safety*, North-Holland Publishing Co., Amsterdam, 1982.

Jones-Lee, Michael, "The Economic Value of Life: A Comment," *Economica* 54: 397-400, August 1987.

Jones-Lee, Michael, "The Value of Changes in the Probability of Death or Injury," *Journal of Political Economy* 82: 835-849, 1974.

Jouzel, J., C. Lorius, J.R. Petit, C. Genthon, N.I. Barkov, V.M. Kotlyakov, and V.M. Petrov, "Vostok Ice Core: A Continuous Isotope Temperature Record Over the Last Climatic Cycle (160,000 years)," *Nature* 329: 403-408, 1987.

Jungermann, Helmut, Holger Schutz, and Manfred Thuring, "Mental Models in Risk Assessment: Informing People About Drugs," *Risk Analysis* 8: 147-155, March 1988.

Just, Richard E., Darrell L. Hueth, and Andrew Schmitz, *Applied Welfare Economics and Public Policy*, Prentice-Hall, Inc., Englewood Cliffs, New Jersey, 1982.

Kahan, James P., "How Psychologists Talk About Risk," P-6403, The RAND Corporation, Santa Monica, October 1979.

Kahneman, Daniel, and Amos Tversky, "Prospect Theory: An Analysis of Decision Under Risk," *Econometrica* 47 (2), March 1979.

Kahneman, Daniel, Paul Slovic, and Amos Tversky, *Judgment Under Uncertainty: Heuristics and Biases*, Cambridge University Press, Cambridge, 1982.

Kanouse, David E., and Barbara Hayes-Roth, "Cognitive Considerations in the Design of Product Warnings," *Banbury Report 6: Product Labeling and Health Risks*, Cold Spring Harbor Laboratory, New York, 1980.

Kanouse, David E., Sandra H. Berry, Barbara Hayes-Roth, William H. Rogers, and John D. Winkler, *Informing Patients About Drugs: Summary Report on Alternative Designs for Prescription Drug Leaflets*, R-2800-FDA, The RAND Corporation, Santa Monica, August 1981.

Karl, Thomas R., J. Dan Tarpley, Robert G. Quayle, Henry F. Diaz, David A. Robinson, Raymond S. Bradley, "The Recent Climate Record: What It Can and Cannot Tell Us," *Reviews of Geophysics* 27: 405-430, 1989.

Karni, Edi, and Zvi Safra, "'Preference Reversal' and the Observability of Preferences by Experimental Methods," *Econometrica* 55: 675-685, 1987.

Kealy, Mary Jo, John F. Dovidio, and Mark L. Rockel, "Accuracy in Valuation is a Matter of Degree," *Land Economics* 64: 158-171, 1988.

Keeler, Emmett B., and Shan Cretin, "Discounting of Life-Saving and Other Nonmonetary Effects," *Management Science* 29 (3), March 1983.

Keeney, Ralph L., and Howard Raiffa, *Decisions with Multiple Objectives: Preferences and Value Tradeoffs*, John Wiley and Sons, New York, 1976.

Keeney, Ralph L., and Detlof von Winterfeldt, "Improving Risk Communication," *Risk Analysis* 6: 417-424, 1986.

Keller, Kevin Lane, and Richard Staelin, "Effects of Quality and Quantity of Information on Decision Effectiveness," *Journal of Consumer Research* 14: 200-213, September 1987.

Kelman, Steve, "Cost-Benefit Analysis: An Ethical Critique," *Regulation* 5: 33-40, January/February 1981.

Keown, Charles F., "Risk Judgments and Intention Measures After Reading About Prescription Drug Side Effects in the Format of a Patient Package Insert," *Journal of Consumer Affairs* 17: 277-289, 1983.

Kerr, Richard A., "Is the Greenhouse Here?" *Science* 239: 559-561, 1988.

Khalil, M.A.K., and R.A. Rasmussen, "The Release of Trichlorofluoromethane from Rigid Polyurethane Foams," *Journal of the Air Pollution Control Association* 36, 159-163, 1986.

Kiehl, J.T. Byron A. Boville, and Bruce P. Brieglev, "Response of a General Circulation Model to a Prescribed Antarctic Ozone Hole," *Nature* 332: 501-504, 1988.

Kirchner, James W., "The Gaia Hypothesis: Can It Be Tested?" *Reviews of Geophysics* 27: 223-235, 1989.

Knetsch, Jack L., "Environmental Policy Implications of Disparities between Willingness to Pay and Compensation Demanded Measures of Values," *Journal of Environmental Economics and Management* 18: 227-237, 1990.

Knetsch, Jack L., "The Endowment Effect and Evidence of Nonreversible Indifference Curves," *American Economic Review* 79: 1277-1284, 1989.

Knetsch, Jack L., and J.A. Sinden, "Willingness to Pay and Compensation Demanded: Experimental Evidence of an Unexpected Disparity in Measures of Value," *Quarterly Journal of Economics* 99: 507-521, 1984.

Knez, Peter, Vernon L. Smith, and Arlington W. Williams, "Individual Rationality, Market Rationality, and Value Estimation," *American Economic Review* 75 (2), May 1985.

Kohler, Daniel F., John Haaga, and Frank Camm, *Projections of Consumption of Products Using Chlorofluorocarbons in Developing Countries*, N-2458-EPA, The RAND Corporation, Santa Monica, January 1987.

Kokoski, Mary F., and V. Kerry Smith, "A General Equilibrium Analysis of Partial-Equilibrium Welfare Measures: The Case of Climate Change," *American Economic Review* 77: 331-341, 1987.

Kolb, Jeffrey A., and Joel D. Scheraga, "Discounting the Benefits and Costs of Environmental Regulations," *Journal of Policy Analysis and Management* 9: 381-390, 1990.

Kopp, Raymond J., and V. Kerry Smith, "Benefit Estimation Goes to Court: The Case of Natural Resource Damage Assessments," *Journal of Policy Analysis and Management* 8: 593-612, 1989.

Krewski, D., K.S. Crump, J. Farmer, D.W. Gaylor, R. Howe, C. Portier, D. Salsburg, R.L. Sielken, and J. Van Ryzin, "A Comparison of Statistical Methods for Low Dose Extrapolation Utilizing Time-to-Tumor Data," *Fundamental and Applied Toxicology* 3, 1983.

Kuo, Cynthia, Craig Lindberg, and David J. Thomson, "Coherence Established between Atmospheric Carbon Dioxide and Global Temperature," *Nature* 343: 709-713, 1990.

Lancaster, Kelvin, "A New Approach to Consumer Theory," *Journal of Political Economy* 74, 1966.

Landefeld, J. Steven, and Eugene P. Seskin, "The Economic Value of Life: Linking Theory to Practice," *American Journal of Public Health* 72, 1982.

Lashof, D.A., D.R. Ahuja, "Relative Contributions of Greenhouse Gas Emissions to Global Warming," *Nature* 344: 529-531, 1990.

Latin, Howard, "Good Science, Bad Regulation, and Toxic Risk Assessment," *Yale Journal on Regulation* 5: 89-148, Winter 1988.

Lave, Lester B., "Health and Safety Risk Analyses: Information for Better Decisions," *Science* 236: 291-295, 1987.

Lave, Lester B., "Mitigating Strategies for Carbon Dioxide Problems," *American Economic Review—Proceedings* 72: 257-261, May 1982.

Lave, Lester B., "The Greenhouse Effect: What Government Actions are Needed? *Journal of Policy Analysis and Management* 7: 460-470, 1988.

Lave, Lester B., and Gilbert S. Omenn, "Cost-Effectiveness of Short-Term Tests for Carcinogenicity," *Nature* 324: 29-34, 1986.

Lave, L., and K.H. Vickland, "Adjusting to Greenhouse Effects: The Demise of Traditional Cultures and the Cost to the USA," *Risk Analysis* 9:283-291, 1989.

Leaf, Alexander, "Potential Health Effects of Global Climatic and Environmental Changes," *New England Journal of Medicine* 321: 1577-1583, 1989.

Leamer, Edward E., *Specification Searches*, John Wiley and Sons, New York, 1978.

Lee, Terence R., "Public Attitudes Towards Chemical Hazards," *The Science of the Total Environment* 51, 125-147, 1986.

Leigh, J. Paul, "Schooling and Seat Belt Use," *Southern Economic Journal* 57: 195-207, 1990.

Lenard, Thomas M., and Michael P. Mazur, "Harvest of Waste: The Marketing Order Program," *Regulation*, May/June 1985.

Leonard, Herman B., "Elicitation of Honest Preferences for the Assignment of Individuals to Positions," *Journal of Political Economy* 91: 461-479, 1983.

Leontief, Wassily, "Population Growth and Economic Development: Illustrative Projections," *Population and Development Review* 5 (1), 1979.

Levine, A.S., T.P. Labuza, and J.E. Morley, "Food Technology: A Primer for Physicians," *New England Journal of Medicine* 312, 1985.

Levy, Alan S., Odonna Mathews, Marilyn Stephenson, Janet E. Tenney, and Raymond E. Schucker, "The Impact of a Nutrition Information Program on Food Purchases," *Journal of Public Policy and Marketing* 4: 1-13, 1985.

Levy, Stuart B., "Antibiotic Resistant Bacteria in Food of Man and Animals," in Malcolm Woodbine, ed., *Antimicrobials in Agriculture*, Butterworth and Co., Ltd., London, 1984.

Lichtenberg, Erik, and David Zilberman, "Efficient Regulation of Environmental Health Risks," *Quarterly Journal of Economics* 103: 167-178, 1988.

Lind, Robert C., Kenneth J. Arrow, Gordon R. Corey, Partha Dasgupta, Amartya K. Sen, Thomas Stauffer, Joseph E. Stiglitz, J.A. Stockfish, and Robert Wilson, *Discounting for Time and Risk in Energy Policy*, Resources for the Future, Washington, D.C., 1982.

Lindgren, Georg, "Model Processes in Nonlinear Prediction with Application to Detection and Alarm," *The Annals of Probability* 8: 775-792, 1980.

Lindley, D.V., "The Probability Approach to the Treatment of Uncertainty in Artificial Intelligence and Expert Systems," *Statistical Science* 2: 17-24, February 1987.

Lindsay, D.G., and J.C. Sherlock, "Environmental Contaminants," in E. F. Patrice Jellife and Derrick B. Jellife, eds., *Adverse Effects of Foods*, Plenum Press, New York, 1982.

Linnerooth, Joanne, "The Value of Human Life: A Review of the Models," *Economic Inquiry* 17, 1979.

Loomes, Graham, and Robert Sugden, "Disappointment and Dynamic Consistency in Choice under Uncertainty," *Review of Economic Studies* 53: 271-282, 1986.

Lorenz, E.N., "The Predictability of a Flow which Possesses Many Scales of Motion," *Tellus* 21: 289-307, 1969a.

Lorenz, E.N., "Three Approaches to Atmospheric Predictability," *Bulletin of the American Meteorological Society* 50: 345-349, 1969b.

Lovelock, J.E., "Atmospheric Halocarbons and Stratospheric Ozone," *Nature* 252: 292-294, 1974.

Lovelock, James E., "Geophysiology, The Science of Gaia," *Reviews of Geophysics* 27: 215-222, 1989.

Lovins, Amory B., "Energy Strategy: The Road Not Taken?" *Foreign Affairs* 55: 61-96, 1976.

Lovins, A.B., L.H. Lovins, F. Krause, and W. Bach, *Least Cost Energy: Solving the CO_2 Problem*, Brick House, Andover, MA, 1981.

Lowe, David C., Carl A.M. Brenninkmeijer, Martin R. Manning, Rodger Sparks, and Gavin Wallace, "Radiocarbon Determination of Atmospheric Methane at Baring Head, New Zealand," *Nature* 332: 522-524, 1988.

Lucas, Robert E.B., "Hedonic Wage Equations and Psychic Wages in the Returns to Schooling," *American Economic Review* 67, 1977.

Lund, Adrian K., "Voluntary Seat Belt Use Among U.S. Drivers: Geographic, Socioeconomic, and Demographic Variation," *Accident Analysis and Prevention* 18: 43-50, 1986.

MacDonald, Gordon J., "Scientific Basis for the Greenhouse Effect," *Journal of Policy Analysis and Management* 7: 425-444, 1988.

MacFadyen, J. Tevere, "Behind the Natural-Foods Facade," *B and K's Country Journal*, August, 1984.

Machina, Mark J., "'Expected Utility' Analysis Without the Independence Axiom," *Econometrica* 50 (2), March 1982.

Machina, Mark J., "Decision-Making in the Presence of Risk," *Science*, 236: 537-543, 1987.

Machina, Mark J., "Temporal Risk and the Nature of Induced Preferences," *Journal of Economic Theory* 33, 1984.

Machina, Mark J., *The Economic Theory of Individual Behavior Toward Risk: Theory, Evidence and New Directions*, Technical Report No. 433, Institute for Mathematical Studies in the Social Sciences, Stanford University, Stanford, California, October 1983.

MacRae, Duncan Jr., and Dale Whittington, "Assessing Preferences in Cost-Benefit Analysis: Reflections on Rural Water Supply Evaluation in Haiti," *Journal of Policy Analysis and Management* 7: 246-263, 1988.

Manne, Alan S., and Richard G. Richels, "CO_2 Emission Limits: An Economic Cost Analysis for the USA," manuscript, November 1989.

Manne, Alan S., and Richard G. Richels, "Global CO_2 Emission Reductions—The Impacts of Rising Energy Costs," manuscript, February 1990.

Manzer, L.E., "The CFC-Ozone Issue: Progress on the Development of Alternatives to CFCs," *Science* 249: 31-35, 1990.

Marin, Alan, and George Psacharopoulos, "The Reward for Risk in the Labor Market: Evidence from the United Kingdom and a Reconciliation with Other Studies," *Journal of Political Economy* 90 (4), 1982.

Marshall, William E., "Health Foods, Organic Foods, Natural Foods: What They Are and What Makes Them Attractive to Consumers," *Food Technology* 28 (2), February 1974.

Martin, Robert E., "On Judging Quality by Price: Price Dependent Expectations, Not Price Dependent Preferences," *Southern Economic Journal*, January 1986.

Mattos de Lemos, Haroldo, "Amazonia: In Defense of Brazil's Sovereignty," *The Fletcher Forum of World Affairs* 14: 301-312, 1990.

Mayas, J.M.B., N.K. Boyd, M.A. Collins, and B.I. Harris, *A Study of Demographic, Situational, and Motivational Factors Affecting Restraint Usage in Automobiles*, Lawrence Johnson and Associates, Inc., National Highway Traffic Safety Administration, U.S. Department of Transportation publication HS-806-402, Washington, D.C., 1983.

McBean, Lois D., and Elwoos W. Speckmann, "Diet, Nutrition and Cancer," in E.F. Patrice Jellife and Derrick B. Jellife, eds., *Adverse Effects of Foods*, Plenum Press, New York, 1982.

McElroy, Michael B., and Ross J. Salawitch, "Changing Composition of the Global Stratosphere," *Science* 243: 763-770, 1989.

McElveen, James C., Jr., and Pamela S. Eddy, "Cancer and Toxic Substances: The Problem of Causation and the Use of Epidemiology," *Personal Injury Deskbook*, Matthew Bender and Co., New York, 1985.

McEwen, F.L., "Food Production—The Challenge for Pesticides," *Bioscience* 28, 1978.

McKibben, Bill, *The End of Nature*, Random House, New York, 1989.

McLaren, A.S., R.G. Barry, and R.H. Bourke, "Could Arctic Ice be Thinning?" *Nature* 345: 762, 1990.

McNeil, Barbara J., Ralph Weichselbaum, and Stephen G. Pauker, "Fallacy of the Five-Year Survival in Lung Cancer," *New England Journal of Medicine* 299, 1978.

Melburg, Valerie, and James T. Tedeschi, "Risk-Taking, Justifiability, Recklessness, and Responsibility," *Personality and Social Psychology Bulletin* 7: 509-515, 1981.

Mendeloff, John M., and Robert M. Kaplan, "Are Large Differences in 'Lifesaving' Costs Justified? A Psychometric Study of the Relative Value Placed on Preventing Deaths," *Risk Analysis* 9: 349-363, 1989.

Mendeloff, John M., *The Dilemma of Toxic Substance Regulation: How Overregulation Causes Underregulation at OSHA*, MIT Press, Cambridge, 1988.

Merrill, Richard A., "FDA's Implementation of the Delaney Clause: Repudiation of Congressional Choice or Reasoned Adaptation to Scientific Progress?" *Yale Journal on Regulation* 5: 1-88, Winter 1988.

Mikolajewicz, Uwe, Benjamin D. Santer, and Ernst Maier-Reimer, "Ocean Responses to Greenhouse Warming," *Nature* 345: 589-593, 1990.

Miller, Jon R., and Frank Lad, "Flexibility, Learning, and Irreversibility in Environmental Decisions: A Bayesian Approach," *Journal of Environmental Economics and Management* 11 (2), 1984.

Miller, Ted R., *Benefit-Cost Analyses of Health and Safety: Conceptual and Empirical Issues*, Project Report, The Urban Institute, Washington, D.C., December 1986.

Milon, J. Walter, Jonathan Gressel, and David Mulkey, "Hedonic Amenity Valuation and Functional Form Specification," *Land Economics* 60 (4), November 1984.

Milvy, Paul, "A General Guideline for Management of Risk from Carcinogens," *Risk Analysis* 6: 69-79, March 1986.

Mintzer, Irving, "Living in a Warmer World: Challenges for Policy Analysis and Management," *Journal of Policy Analysis and Management* 7: 445-459, 1988.

Mintzer, Irving, *A Matter of Degrees: The Potential for Limiting the Greenhouse Effect*, World Resources Institute, Washington, D.C., 1987.

Mishan, E.J., *Cost-Benefit Analysis*, Praeger, New York, 1971.

Mitchell, Bridger M., and James R. Vernon, *The Health Costs of Skin Cancer Caused by Ultraviolet Radiation*, N-2538-EPA, The RAND Corporation, Santa Monica, February 1987.

Mitchell, John F.B., "The 'Greenhouse' Effect and Climate Change," *Reviews of Geophysics* 27: 115-139, 1989.

Mitchell, Robert C., and Richard T. Carson, *Using Surveys to Value Public Goods: The Contingent Valuation Method*, Resources for the Future, Washington, D.C., 1989.

Mitchell, Robert C., and Richard T. Carson, *Valuing Drinking Water Risk Reductions using the Contingent Valuation Method: A Methodological Study of Risks from THM and Giardia*, Resources for the Future, Washington, D.C., 1986.

Mitchell, Robert Cameron, and Richard T. Carson, "Validity and Reliability in Contingent Valuation Surveys," Draft Working Paper, Resources for the Future, Washington, D.C., August 22, 1985.

Molina, Mario J., and F. S. Rowland, "Stratospheric Sink for Chlorofluoromethanes: Chlorine Atom-Catalysed Destruction of Ozone," *Nature* 249: 810-812, 1974.

Molina, Mario J., Tai-Ly Tso, Luisa T. Molina, and Franc C.-Y. Wang, "Antarctic Stratospheric Chemistry of Chlorine Nitrate, Hydrogen Chloride, and Ice: Release of Active Chlorine," *Science* 238: 1253-1257, 1987.

Montgomery, John D., "Environmental Management as a Third-World Problem," *Policy Sciences* 23: 163-176, 1990.

Moomaw, William R., "Scientific and International Policy Responses to Global Climate Change," *The Fletcher Forum of World Affairs* 14: 249-261, 1990.

Moore, Michael J., and W. Kip Viscusi, "Doubling the Estimated Value of Life: Results Using New Occupational Fatality Data," *Journal of Policy Analysis and Management* 7: 476-490, 1988.

Mooz, William E., and T.H. Quinn, *Flexible Urethane Foams and Chlorofluorocarbon Emissions*, N-1472-EPA, The RAND Corporation, Santa Monica, June 1980.

Mooz, W.E., S.H. Dole, D.L. Jaquette, W.H. Krase, P.F. Morrison, S.L. Salem, R.G. Salter and K.A. Wolf, *Technical Options for Reducing Chlorofluorocarbon Emissions*, R-2879-EPA, The RAND Corporation, Santa Monica, March 1982.

Mooz, William E., Kathleen A. Wolf, and Frank Camm, *Potential Constraints on Cumulative Global Production of Chlorofluorocarbons*, R-3400-EPA, The RAND Corporation, Santa Monica, May 1986.

Morgan, M. Granger, Paul Slovic, Indira Nair, Dan Geisler, Donald MacGregor, Baruch Fischhoff, David Lincoln, and Keith Florig, "Powerline Frequency Electric and Magnetic Fields: A Pilot Study of Risk Perception," *Risk Analysis* 5 (2), June 1985.

Morgan, M. Granger, Samuel C. Morris, Max Henrion, Deborah A.L. Amaral, and William R. Rish, "Technical Uncertainty in Quantitative Policy Analysis—A Sulfur Air Pollution Example," with comments, *Risk Analysis* 4: 201-216, 1984.

Morrall, John F. III, "A Review of the Record," *Regulation* 25-34, November/December 1986.

Moses, Marion, "Pesticides," in William N. Rom, ed., *Environmental and Occupational Medicine*, Little, Brown and Co., Boston, 1983.

Mott, Lawrie, *Pesticides in Food: What the Public Needs to Know*, Natural Resources Defense Council, San Francisco, 1984.

Mueller, Dennis C., *Public Choice*, Cambridge University Press, Cambridge, 1979.

Mumpower, Jeryl, "An Analysis of the *de minimis* Strategy for Risk Management," *Risk Analysis* 6: 437-446, 1986.

Murray, Michael P., "Mythical Demands and Mythical Supplies for Proper Estimation of Rosen's Hedonic Price Model," *Journal of Urban Economics* 14, 1983.

Mustafa, Mohammad G., "Agricultural Chemicals," in E.F. Patrice Jellife and Derrick B. Jellife, eds., *Adverse Effects of Foods*, Plenum Press, New York, 1982.

National Academy of Sciences, *Causes and Effects of Changes in Stratospheric Ozone: Update 1983*, Washington D.C., 1984.

National Academy of Sciences, *Causes and Effects of Stratospheric Ozone Depletion: An Update*, Washington D.C., 1982.

National Academy of Sciences, *Halocarbons: Effects on Stratospheric Ozone*, Washington D.C., 1976.

National Academy of Sciences, *Protection against Depletion of Stratospheric Ozone*, Washington D.C., 1979.

National Institute for Occupational Safety and Health (NIOSH), *Registry of Toxic Effects of Chemical Substances*, U.S. Department of Health and Human Services, Washington, D.C., October 1984.

National Research Council, *Changing Climate: Report of the Carbon Dioxide Assessment Committee*, National Academy Press, Washington, D.C., 1983.

National Research Council, *Diet, Nutrition, and Cancer*, National Academy Press, Washington, D.C., 1982.

National Research Council, *Pharmacokinetics in Risk Assessment: Drinking Water and Health, Volume 8*, National Academy Press, Washington, D.C., 1987.

National Research Council, *Regulating Pesticides in Food: The Delaney Paradox*, National Academy Press, Washington, D.C., 1987.

National Research Council, *Regulating Pesticides*, National Academy of Sciences, Washington D.C., 1980.

National Research Council, *Risk Assessment in the Federal Government: Managing the Process*, National Academy Press, Washington, D.C., 1983.

National Research Council, *Toxicity Testing: Strategies to Determine Needs and Priorities*, National Academy of Sciences, 1984.

Nordhaus, William D., "The Economics of the Greenhouse Effect," manuscript, June 1989.

Nordhaus, William D., and Gary W. Yohe, "Future Paths of Energy and Carbon Dioxide Emissions," in National Research Council, *Changing Climate: Report of the Carbon Dioxide Assessment Committee*, National Academy Press, Washington D.C., 1983.

Norton, Bryan G., ed., *The Preservation of Species: The Value of Biological Diversity*, Princeton University Press, Princeton, N.J., 1986.

O'Day, James, and Robert E. Scott, "Seatbelt Use, Ejection and Entrapment," *Health Education Quarterly* 11 (2), 1984.

O'Neill, Brian, Allan F. Williams, and Ronald S. Karpf, "Passenger Car Size and Driver Seatbelt Use, *American Journal of Public Health* 73: 588-590, 1983.

Office of Technology Assessment (U.S. Congress), Oceans and Environment Program, *An Analysis of the Montreal Protocol on Substances that Deplete the Ozone Layer*, revised February 1, 1988.

Ohta, Makoto, and Zvi Griliches, "Automobile Prices and Quality: Did the Gasoline Price Increases Change Consumer Tastes in the U.S.?" *Journal of Business and Economics Statistics* 4 (2), April 1986.

Ohta, Makoto, and Zvi Griliches, "Automobile Prices Revisited: Extensions of the Hedonic Hypothesis," in N. Terleckyj, ed., *Household Production and Consumption*, National Bureau of Economic Research, New York, 1976.

Oi, Walter Y., "The Economics of Product Safety," *Bell Journal of Economics and Management Science* 4 (1), Spring 1973.

Okrent, David, "The Safety Goals of the U.S. Nuclear Regulatory Commission," *Science* 236: 296-300, 1987.

Olcott, H.S., "What is Known about Human Nutritional Need," *American Journal of Agricultural Economics* 60, December 1978.

Oppenheimer, Michael, and Robert H. Boyle, *Dead Heat: The Race Against the Greenhouse Effect*, Basic Books, Inc., New York, 1990.

Organically Grown Foods: A General Handbook, Rodale Press, Emmaus, Pennsylvania, 1971.

Ott, Wayne R., *Environmental Indices: Theory and Practice*, Ann Arbor Science Publishers, Inc., Ann Arbor, Michigan, 1978.

Page, Talbot, "A Generic View of Toxic Chemicals and Similar Risks," *Ecology Law Quarterly* 7: 207-244, 1978.

Palmer, Adele R., and Timothy H. Quinn, *Allocating Chlorofluorocarbon Permits: Who Gains, Who Loses, and What Is the Cost?*, R-2806-EPA, The RAND Corporation, Santa Monica, July 1981.

Palmer, Adele R., and Timothy H. Quinn, *Economic Impact Assessment of a Chlorofluorocarbon Production Cap*, N-1656-EPA, The RAND Corporation, Santa Monica, February 1981.

Palmer, Adele R., William E. Mooz, Timothy H. Quinn and Kathleen A. Wolf, *Economic Implications of Regulating Chlorofluorocarbon Emissions from Nonaerosol Applications*, R-2524-EPA, The RAND Corporation, Santa Monica, June 1980.

Park, Colin N., and Ronald D. Snee, "Quantitative Risk Assessment: State-of-the-Art for Carcinogenesis," *The American Statistician* 37: 427-441, November 1983.

Pastor, John, and W.M. Post, "Response of Northern Forests to CO_2-Induced Climate Change," *Nature* 334: 55-58, 1988.

Pate-Cornell, M. Elisabeth, "Fault Trees vs. Event Trees in Reliability Analysis," *Risk Analysis* 4: 177-186, 1984.

Pate-Cornell, M. Elisabeth, "Warning Systems in Risk Management," *Risk Analysis* 6: 223-234, 1986.

Pate-Cornell, M. Elisabeth, H.L. Lee, and G. Tagaras, "Warnings of Malfunction: The Decision to Inspect and Maintain Production Processes On Schedule or On Demand," *Management Science* 33: 1277-1290, October 1987.

Payne, Melanie S., "Preparing for Group Interviews," *Advances in Consumer Research* 4, 1976.

Pearman, G.I., and P.J. Fraser, "Sources of Increased Methane," *Nature* 332: 449-450, 1988.

Peck, Stephen C., and Richard G. Richels, "The Value of Information to the Acidic Deposition Debates," *Journal of Business and Economic Statistics* 5: 205-217, 1987.

Pedersen, David H., and Richard W. Hornung, "Computerized Estimates of Potential Occupational Health Risk Due to Chemical Exposure," *Risk Analysis* 6 (1), March 1986.

Pedersen, David H., Randy O. Young, and David S. Sundin, *A Model for the Identification of High Risk Occupational Groups Using RTECS and NOHS Data*, NIOSH Technical Report, U.S. Department of Health and Human Services, Cincinnati, 1983.

Peers, F.G., and C.A. Linsell, "Dietary Aflatoxins and Liver Cancer—A Population Based Study in Kenya," *British Journal of Cancer* 27, 1973.

Peltzman, Sam, "The Effects of Automobile Safety Regulation," *Journal of Political Economy* 83: 677-725, 1975a.

Peltzman, Sam, *Regulation of Automobile Safety*, American Enterprise Institute, Washington, D.C., 1975b.

Penner, Joyce E., "Cloud Albedo, Greenhouse Effects, Atmospheric Chemistry, and Climate Change," *Journal of the Air and Waste Management Association* 40: 456-461, 1990.

Peters, Charles M., Alwyn H. Gentry, and Robert O. Mendelsohn, "Valuation of an Amazonian Rainforest," *Nature* 339: 655-656, 1989.

Peto, Richard, Malcolm C. Pike, Leslie Bernstein, Lois Swirsky Gold, and Bruce N. Ames, "The TD_{50}: A Proposed General Convention for the Numerical Description of the Carcinogenic Potency of Chemicals in Chronic-Exposure Animal Experiments," *Environmental Health Perspectives* 58, 1984.

Pimentel, David, John Krummel, David Gallahan, Judy Hough, Alfred Merrill, Ilse Schreiner, Pat Vittum, Fred Koziol, Ephraim Back, Doreen Yen, and Sandy Fiance, "Benefits and Costs of Pesticide Use in U.S. Food Production," *Bioscience* 28 (12), 1978.

Prather, Michael J., Michael B. McElroy, and Steven C. Wofsy, "Reductions in Ozone at High Concentrations of Stratospheric Halogens," *Nature* 312: 227-231, 1984.

Pratt, John W., Howard Raiffa, and Robert Schlaifer, *Introduction to Statistical Decision Theory*, prelim. ed., McGraw-Hill Book Company, New York, 1965.

Quinn, Timothy, Kathleen A. Wolf, William E. Mooz, James K. Hammitt, Thomas W. Chesnutt and Syam Sarma, *Projected Use, Emissions and Banks of Potential Ozone Depleting Substances*, N-2282-EPA, The RAND Corporation, Santa Monica, January 1986.

Quirk, James P., and Katsuaki L. Terasawa, *The Choice of Discount Rate Applicable to Government Resource Use*, R-3464-PA&E, The RAND Corporation, Santa Monica, December 1987.

Raiffa, Howard, *Decision Analysis: Introductory Lectures on Choices under Uncertainty*, Addison-Wesley Publishing Company, Reading, Massachusetts, 1968.

Ramanathan, V., "Greenhouse Effect Due to Chlorofluorocarbons: Climatic Implications," *Science* 190: 50-51, 1975.

Ramanathan, V., "The Greenhouse Theory of Climate Change: A Test by an Inadvertent Global Experiment," *Science* 240: 293-299, 1988.

Ramanathan, V., R.D. Cess, E.F. Harrison, P. Minnis, B.R. Barkstrom, E. Ahmad, D. Hartmann, "Cloud-Radiative Forcing and Climate: Results from the Earth Radiation Budget Experiment," *Science* 243: 57-63, 1989.

Ramanathan, V., R.J. Cicerone, H.B. Singh, and T.J. Kiehl, "Trace Gases and Their Potential Role in Climatic Change," *Journal of Geophysical Research* 90 (D3): 5547-5566, 1985.

Rasmussen, R.A., and M.A.K. Khalil, "Atmospheric Trace Gases: Trends and Distributions Over the Last Decade," *Science* 232, 1623-1624, 27 June 1986.

Raval, A., and V. Ramanathan, "Observational Determination of the Greenhouse Effect," *Nature* 342: 758-761, 1990.

Raynaud, D., J. Chappellaz, J.M. Barnola, Y.S. Korotkevich, and C. Lorius, "Climatic and CH_4 Cycle Implications of Glacial-Interglacial CH_4 Change in the Vostok Ice Core," *Nature* 333: 655-657, 1988.

Reiling, Stephen D., Kevin J. Boyle, Marcia L. Phillips, and Mark W. Anderson, "Temporal Reliability of Contingent Values," *Land Economics* 66: 128-134, 1990.

Reinsel, Gregory C., and George C. Tiao, "Impact of Chlorofluoromethanes on Stratospheric Ozone: A Statistical Analysis of Ozone Data for Trends," *Journal of the American Statistical Association* 82: 20-30, 1987.

Reinsel, G., G.C. Tiao, M.N. Wang, R. Lewis, and D. Nychka, "Statistical Analysis of Stratospheric Ozone Data for the Detection of Trends," *Atmospheric Environment* 15: 1569-1577, 1981.

Reinsel, Gregory C., George C. Tiao, John J. DeLuisi, Carl L. Mateer, Alvin J. Miller, and John E. Frederick, "Analysis of Upper Stratosphere Umkehr Ozone Profile Data for Trends and the Effects of Stratospheric Aerosols," *Journal of Geophysical Research* 89 (D3): 4833-4840, 1984.

Revelle, Roger R., and Paul E. Waggoner, "Effects of a Carbon Dioxide-Induced Climatic Change on Water Supplies in the Western United States," in *Changing Climate: Report of the Carbon Dioxide Assessment Committee*, National Academy Press, Washington, D.C., 1983.

Reynolds, Steven H., Shari J. Stowers, Rachel M. Patterson, Robert R. Maronpot, Stuart A. Aaronson, and Marshall W. Anderson, "Activated Oncogenes in B6C3F1 Mouse Liver Tumors: Implications for Risk Assessment," *Science* 237, 1309-1316, 11 September 1987.

Roberts, Gail, "Raising the Organic Standard," *Harrowsmith*, Camden House Publishing, Ltd., Camden East, Ontario, Canada, February/March 1985.

Roberts, Tanya, "Microbial Pathogens in Raw Pork, Chicken, and Beef: Benefit Estimates for Control using Irradiation," *American Journal of Agricultural Economics* 67: 957-965, 1985.

Rodhe, Henning, "A Comparison of the Contribution of Various Gases to the Greenhouse Effect," *Science* 248: 1217-1219, 1990.

Rosen, Sherwin, "Hedonic Prices and Implicit Markets: Product Differentiation in Pure Competition," *Journal of Political Economy* 82, 1974.

Rosen, Sherwin, "The Value of Changes in Life Expectancy," *Journal of Risk and Uncertainty* 1: 285-304, 1988.

Rosen, Sherwin, "Valuing Health Risk," *American Economic Review* 71: 241-245, 1981.

Rosenberg, N.J., W.E. Easterling III, P.R. Crosson, J. Darmstadter, eds., *Greenhouse Warming: Abatement and Adaptation*, Resources for the Future, Washington, D.C., 1989.

Ross, Steven S., ed., *Toxic Substances Sourcebook*, Environment Information Center, Inc., New York, 1978.

Rothschild, Michael, "Searching for the Lowest Price when the Distribution of Prices is Unknown," *Journal of Political Economy*, 82: 689-711, 1974.

Russell, Milton, and Michael Gruber, "Risk Assessment in Environmental Policy Making," *Science* 236: 286-290, 1987.

Russo, J. Edward, Richard Staelin, Catherine A. Nolan, Gary J. Russell, and Barbara L. Metcalf, "Nutrition Information in the Supermarket," *Journal of Consumer Research* 13, June 1986.

Satterthwaite, M.A., "Strategy Proofness and Arrow's Conditions: Existence and Correspondence Theorems for Voting Procedures and Social Welfare Functions," *Journal of Economic Theory* 10, April 1975.

Schaefer, Bradley E., "The Astrophysics of Suntanning," *Sky and Telescope* 595-596, June 1988.

Schelling, Thomas C., "Climatic Change: Implications for Welfare and Policy," in *Changing Climate: Report of the Carbon Dioxide Assessment Committee*, National Academy Press, Washington, D.C., 1983.

Schelling, Thomas C., "The Life You Save May Be Your Own," in Samuel B. Chase, Jr., ed., *Problems in Public Expenditure Analysis*, Brookings Institution, Washington D.C., 1968.

Schenck, Anna P., Carol W. Runyan, and Jo Anne L. Earp, "Seat Belt Use Laws: The Influence of Data on Public Opinion," *Health Education Quarterly* 12 (4), 1985.

Schlesinger, Michael E., "Climate Model Simulations of CO_2-Induced Climate Change," *Advances in Geophysics* 26: 141-235, 1984.

Schlesinger, Michael E., and John F.B. Mitchell, "Climate Model Simulations of the Equilibrium Climatic Response to Increased Carbon Dioxide," *Reviews of Geophysics* 24: 760-798, 1987.

Schmucker, K.J., *Fuzzy Sets, Natural Language, Computations, and Risk Analysis*, Computer Science Press, Rockville, Maryland, 1984.

Schneider, Stephen H., "The Greenhouse Effect: Science and Policy," *Science* 243: 771-781, 1989a.

Schneider, Stephen H., *Global Warming: Are We Entering The Greenhouse Century?* Sierra Club Books, San Francisco, 1989b.

Schoemaker, Paul J. H., "The Expected Utility Model: Its Variants, Purposes, Evidence and Limitations," *Journal of Economic Literature* 20: 529-563, June 1982.

Schucker, Raymond E., Raymond C. Stokes, Michael L. Stewart, and Douglas P. Henderson, "The Impact of the Saccharin Warning Label on Sales of Diet Soft Drinks in Supermarkets," *Journal of Public Policy and Marketing* 2: 46-56, 1983.

Schutz, Howard G., and Oscar A. Lorenz, "Consumer Preferences for Vegetables Grown Under 'Commercial' and 'Organic' Conditions," *Journal of Food Science* 41 (1), January/February 1976.

Scotchmer, Suzanne, "Hedonic Prices and Cost/Benefit Analysis," *Journal of Economic Theory* 37, 1985.

Scotto, Joseph, Gerald Cotton, Frederick Urbach, Daniel Berger, and Thomas Fears, "Biologically Effective Ultraviolet Radiation: Surface Measurements in the United States. 1974 to 1985," *Science* 239, 762-764, 12 February 1988.

Sedjo, Roger A., and Allen M. Solomon, "Climate and Forests," in N.J. Rosenberg, W.E. Easterling III, P.R. Crosson, and J. Darmstadter, eds., *Greenhouse Warming: Abatement and Adaptation*, Resources for the Future, Washington, D.C., 1989.

Seidel, Stephen, and Dale Keyes, *Can We Delay a Greenhouse Warming?*, U.S. Environmental Protection Agency, EPA-230-10-84-001, Washington, D.C., 1983.

Seinfeld, J.H., "Urban Air Pollution: State of the Science," *Science* 243: 745-752, 1989.

Seller, Christine, John R. Stoll, and Jean-Paul Chavas, "Validation of Empirical Measures of Welfare Change: A Comparison of Nonmarket Techniques," *Land Economics* 61 (2), May 1985.

Shafer, Glen, "Belief Functions and Parametric Models," *Journal of the Royal Statistical Society* (Series B) 44: 322-352, 1982.

Shafer, Glen, *A Mathematical Theory of Evidence*, Princeton University Press, Princeton, New Jersey, 1976.

Shank, R.C., J.E. Gordon, and G.N. Wogan, "Dietary Aflatoxins and Human Liver Cancer III: Field Survey of Rural Thai Families for Ingested Aflatoxins," *Food and Cosmetics Toxicology* 10, 1972.

Sharlin, Harold Issadore, "EDB: A Case Study in Communicating Risk," *Risk Analysis* 6 (1), March 1986.

Shepard, Donald S., and Richard J. Zeckhauser, "Survival versus Consumption," *Management Science* 30, 1984.

Shepard, Donald, and Richard Zeckhauser, "Where Now for Saving Lives?" *Law and Contemporary Problems* 40, 1976.

Shulstad, Robert N., and Herbert H. Stoevener, "The Effects of Mercury Contamination in Pheasants on the Value of Pheasant Hunting in Oregon," *Land Economics* 54 (1), 1978.

Silberberg, Eugene, "Nutrition and the Demand for Tastes," *Journal of Political Economy* 93 (5), 1985.

Sims, John H., and Duane D. Baumann, "Educational Programs and Human Response to Natural Hazards," *Environment and Behavior* 15: 165-189, 1983.

Singer, S. Fred, ed., *Global Climate Change: Human and Natural Influences*, Paragon House, New York, 1989.

Sloss, E.M., and T.P. Rose, *Possible Health Effects of Increased Exposure to Ultraviolet Radiation*, N-2330-EPA, The RAND Corporation, Santa Monica, July 1985.

Slovic, Paul, "Informing and Education the Public About Risk," *Risk Analysis* 6: 403-415, 1986.

Slovic, Paul, "Perception of Risk," *Science* 236, 280-285, April 17, 1987.

Slovic, Paul, Baruch Fischhoff, and Sarah Lichtenstein, "Accident Probabilities and Seat Belt Usage: A Psychological Perspective," *Accident Analysis and Prevention* 10, 1978.

Slovic, Paul, Baruch Fischhoff, Sarah Lichtenstein, Bernard Corrigan, and Barbara Combs, "Preference for Insuring Against Probable Small Losses: Insurance Implications," *Journal of Risk and Insurance* 44, 1977.

Smith, Aileen M., "Testing for Carcinogenicity," *New Engineer*, April 1977, reprinted in Steven S. Ross, ed., *Toxic Substances Sourcebook*, Environment Information Center, Inc., New York, 1978.

Smith, V. Kerry, ed., *Advances in Applied Microeconomics*, Vol. 1, JAI Press, Greenwich, Connecticut, 1981.

Smith, V. Kerry, "Can We Measure the Economic Value of Environmental Amenities?" *Southern Economic Journal* 56: 865-878, 1990.

Smith, V. Kerry, "Nonuse Values in Benefit Cost Analysis," *Southern Economic Journal* 54: 19-26, July 1987.

Smith, V. Kerry, "Supply Uncertainty, Option Price, and Indirect Benefit Estimation," *Land Economics* 61: 303-307, August 1985.

Smith, V. Kerry, and William H. Desvousges, "An Empirical Analysis of the Economic Value of Risk Changes," *Journal of Political Economy* 95: 89-114, 1987.

Smith, V. Kerry, and William H. Desvousges, "Asymmetries in the Valuation of Risk and the Siting of Hazardous Waste Disposal Facilities," *American Economic Review* 76: 291-294, May 1986.

Smith, V. Kerry, and William H. Desvousges, "The Valuation of Environmental Risks and Hazardous Waste Policy," *Land Economics* 64: 211-219, 1988.

Smith, V. Kerry, and William H. Desvousges, "The Value of Avoiding a *LULU*: Hazardous Waste Disposal Sites," *Review of Economics and Statistics* 68 (2), May 1986.

Smith, V. Kerry, and F. Reed Johnson, "How Do Risk Perceptions Respond to Information? The Case of Radon," *Review of Economics and Statistics* 70: 1-8, February 1988.

Smith, V. Kerry, and Yoshiaki Kaoru, "Signals or Noise? Explaining the Variation in Recreation Benefit Estimates," *American Journal of Agricultural Economics* 72: 419-433, 1990.

Smith, V. Kerry, and Yoshiaki Kaoru, "The Hedonic Travel Cost Model: A View from the Trenches," *Land Economics* 63: 179-192, 1987.

Smith, V. Kerry, William H. Desvousges, and A. Myrick Freeman III, *Valuing Changes in Hazardous Waste Risks: A Contingent Valuation Analysis*, Draft Interim Report, Research Triangle Institute, 1985.

Smith, V. Kerry, William F. Desvousges, F. Reed Johnson, and Ann Fisher, "Can Public Information Programs Affect Risk Perceptions?" *Journal of Policy Analysis and Management* 9: 41-59, 1990.

Solomon, Kenneth A., William E. Kastenberg, and Pamela F. Nelson, *Dealing with Uncertainty Arising Out of Probabilistic Risk Assessment*, R-3045-ORNL, The RAND Corporation, Santa Monica, 1983.

Solomon, P., B. Connor, R.L. deZafra, A. Parrish, J. Barrett, and M. Jaramillo, "High Concentrations of Chlorine Monoxide at Low Altitudes in the Antarctic Spring Stratosphere: Secular Variation," *Nature* 328: 411-413, 1987.

Solow, Andrew, and James M. Broadus, "Global Warming: Quo Vadis?" *The Fletcher Forum of World Affairs* 14: 262-269, 1990.

Somerville, Richard C.J., "The Predictability of Weather and Climate," *Climatic Change* 11: 239-246, 1987.

Spencer, R.W., and J.R. Christy, "Precise Monitoring of Global Temperature Trends from Satellites," *Science* 247: 1558-1562, 1990.

Starr, Chauncey, "Social Benefit versus Technological Risk: What is Our Society Willing to Pay for Safety?" *Science* 165, 1969.

Stavins, Robert N., and Adam B. Jaffe, "Unintended Impacts of Public Investments on Private Decisions: The Depletion of Forested Wetlands," *American Economic Review* 80: 337-352, 1990.

Stein, M., "Large Sample Properties of Simulations Using Latin Hypercube Sampling," *Technometrics* 29: 143-151, 1987.

Stewart, Ian, *Does God Play Dice?* Basil Blackwell, Oxford, 1989.

Stigler, George J., and Gary S. Becker, "De Gustibus Non Est Disputandum," *American Economic Review* 67: 76-90, 1977.

Stolarski, Richard S., "Fluorocarbons and Stratospheric Ozone: A Review of Current Knowledge," *The American Statistician* 36: 303-311, 1982.

Stolarski, Richard S., "The Antarctic Ozone Hole," *Scientific American* 258: 30-36, January 1988.

Stolarski, R.S., A.J. Krueger, M.R. Schoeberl, R.D. McPeters, P.A. Newman, and J.C. Alpert, "Nimbus 7 Satellite Measurements of the Springtime Antarctic Ozone Decrease," *Nature* 322: 808-811, 1986.

Stoloff, Leonard, "Carcinogenicity of Aflatoxins," *Science* 237, 1283, 11 September 1987.

Stone, Christopher, "Should Trees Have Standing?—Toward Legal Rights for Natural Objects," *Southern California Law Review* 45: 450-501, 1972.

Stordal, Frode, and Ivar S.A. Isaksen, "Ozone Perturbations Due To Increases in N_2O, CH_4 and Chlorocarbons: Two-Dimensional Time Dependent Calculations," in James G. Titus, ed., *Effects of Changes in Stratospheric Ozone and Global Climate, Vol. I: Overview*, U.S. Environmental Protection Agency and United Nations Environment Programme, Washington, D.C., August 1986.

Surgeon General, *The Surgeon General's Report on Nutrition and Health*, U.S. Department of Health and Human Services, Washington, D.C., 1988.

Swartz, David G., and Ivar E. Strand, Jr., "Avoidance Costs Associated with Imperfect Information: The Case of Kepone," *Land Economics* 57 (2), 1981.

Tans, Pieter P., Inez Y. Fung, Taro Takahashi, "Observational Constraints on the Global Atmospheric CO_2 Budget," *Science* 247: 1431-1438, 1990.

Taylor, Hugh R., Sheila K. West, Frank S. Rosenthal, Beatriz Munoz, Henry S. Newland, Helen Abbey, and Edward A. Emmett, "Effect of Ultraviolet Radiation on Cataract Formation," *New England Journal of Medicine* 319: 1429-1433, 1988.

Tennant, Raymond W., Barry H. Margolin, Michael D. Shelby, Errol Zeiger, Joseph K. Haseman, Judson Spalding, William Caspary, Michael Resnick, Stanley Stasiewicz, Beth Anderson, and Robert Minor, "Prediction of Chemical Carcinogenicity in Rodents from in Vitro Genetic Toxicity Assays," *Science* 236: 933-941, 1987.

Thaler, Richard, "Toward a Positive Theory of Consumer Choice," *Journal of Economic Behavior and Organization* 1, 1980.

Thistle, Paul D., "Psychological Learning Theory and Economic Behavior," *Journal of Behavioral Economics* 13: 67-100, 1984.

Thorslund, Todd W., Charles C. Brown, and Gail Charnley, "Biologically Motivated Cancer Risk Models," *Risk Analysis* 7: 109-119, 1987.

Tiao, G.C., G.C. Reinsel, J.H. Pedrick, G.M. Allenby, C.L. Mateer, A.J. Miller, and J.J. DeLuisi, "A Statistical Trend Analysis of Ozonesonde Data," *Journal of Geophysical Research* 91 (D12): 13121-13136, 1986.

Tideman, J.N., and G. Tullock, "A New and Superior Process for Making Social Choices," *Journal of Political Economy* 84, December 1976.

Titus, James G., ed., *Effects of Changes in Stratospheric Ozone and Climate Change, Vol. I: Overview*, U.S. Environmental Protection Agency and United Nations Environment Programme, Washington, D.C., August 1986a.

Titus, James G., ed., *Effects of Changes in Stratospheric Ozone and Climate Change, Vol. II: Stratospheric Ozone*, U.S. Environmental Protection Agency and United Nations Environment Programme, Washington, D.C., October 1986b.

Titus, James G., ed., *Effects of Changes in Stratospheric Ozone and Climate Change, Vol. III: Climate Change*, U.S. Environmental Protection Agency and United Nations Environment Programme, Washington, D.C., October 1986c.

Titus, James G., ed., *Effects of Changes in Stratospheric Ozone and Climate Change, Vol. IV: Sea Level Rise*, U.S. Environmental Protection Agency and United Nations Environment Programme, Washington, D.C., October 1986d.

Titus, James G., ed., *Greenhouse Effect, Sea Level Rise, and Coastal Wetlands*, EPA-230-05-86-013, U.S. Environmental Protection Agency, Washington, D.C., July 1988.

Tolbert, Margaret A., Michel J. Rossi, Ripudaman Malhotra, and David M. Golden, "Reaction of Chlorine Nitrate with Hydrogen Chloride and Water at Antarctic Stratospheric Temperatures," *Science* 238: 1258-1260, 1987.

Travis, Curtis C., and Robin K. White, "Interspecific Scaling of Toxicity Data," *Risk Analysis* 8: 119-125, March 1988.

Travis, Curtis C., Samantha A. Richter, Edmund A.C. Crouch, Richard Wilson, and Ernest D. Klema, "Cancer Risk Management: A Review of 132 Federal Regulatory Decisions," *Environmental Science and Technology* 21: 415-420, 1987.

Tribe, Laurence H., "Ways Not to Think about Plastic Trees: New Foundations for Environmental Law," *Yale Law Journal* 83: 1315-1348, June 1974.

Triplett, Jack E., "Consumer Demand and Characteristics of Consumption Goods, in N. Terleckyj, ed., *Household Production and Consumption*, National Bureau of Economic Research, New York, 1976.

Trumbull, William N., "Who Has Standing on Cost-Benefit Analysis," *Journal of Policy Analysis and Management* 9: 201-218, 1990.

Tsai, Shan Pou, Eun Sul Lee, and Robert Hardy, "The Effect of a Reduction in Leading Causes of Death: Potential Gains in Life Expectancy," *American Journal of Public Health* 68, 1978.

Tung, Ka-Kit, Maolcolm K.W. Ko, Jose M. Rodriguez, and Nien Dak Sze, "Are Antarctic Ozone Variations a Manifestation of Dynamics or Chemistry?" *Nature* 322: 811-814, 1986.

Tversky, Amos, and Daniel Kahneman, "Judgment under Uncertainty: Heuristics and Biases," *Science* 185, 1974.

Tversky, Amos, and Daniel Kahneman, "Rational Choice and the Framing of Decisions," *Journal of Business* 59: S251-S278, 1986.

Tversky, Amos, and Daniel Kahneman, "The Framing of Decisions and the Psychology of Choice," *Science* 211, January 30, 1981.

U.S. Department of Commerce, The Bureau of the Census, *Statistical Abstract of the United States 1985*, 105th ed., Washington, D.C., 1985.

U.S. Environmental Protection Agency, "Ozone-Depleting Chlorofluorocarbons, Proposed Production Restriction," *Federal Register* 45: 66726-66734, October 7, 1980.

U.S. Environmental Protection Agency, Office of Health and Environmental Assessment, *The Risk Assessment Guidelines of 1986*, EPA/600/8-87/045, Washington, D.C., August 1987; reprinted from *Federal Register* 51: 33992-34054, September 24, 1986.

U.S. Environmental Protection Agency, Office of Health and Environmental Assessment, *Technical Analysis of New Methods and Data Regarding Dichloromethane Hazard Assessments*, External Review Draft, EPA/600/8-87/029A, Washington, D.C., June 1987.

U.S. Environmental Protection Agency, Office of Health and Environmental Assessment, *Update to the Health Assessment Document and Addendum for Dichloromethane (Methylene Chloride): Pharmacokinetics, Mechanism of Action, and Epidemiology*, External Review Draft, EPA/600/8-87/030A, Washington, D.C., July 1987.

U.S. Environmental Protection Agency, Stratospheric Protection Program, Office of Program Development, Office of Air and Radiation, *Regulatory Impact Analysis: Protection of Stratospheric Ozone* (3 volumes), Washington, D.C., December 1987.

U.S. Environmental Protection Agency, "Protection of Stratospheric Ozone: Proposed Rule," *Federal Register* 52: 47489-47523, December 14, 1987.

U.S. Food and Drug Administration, *FDA Compliance Program Report of Findings, FY79 Total Diet Studies—Adult*, No. 7305.002, Washington, D.C., undated.

U.S. Food and Drug Administration, *The FDA Surveillance Index*, NTIS PB82-913299, Washington, D.C., 1981.

U.S. General Accounting Office, *Chemical Data: EPA's Data Collection Practices and Procedures on Chemicals*, GAO/RCED-86-63, Washington, D.C., February 1986.

U.S. General Accounting Office, *Delays and Unresolved Issues Plague New Pesticide Protection Programs*, CED-80-32, Washington, D.C., February 1980.

U.S. General Accounting Office, *Health Risk Analysis: Technical Adequacy in Three Selected Cases*, GAO/PEMD-87-14, Washington, D.C., September 1987.

U.S. General Accounting Office, *Monitoring and Enforcing Food Safety—An Overview of Past Studies*, GAO/RCED-83-153, Washington, D.C., September 1983.

U.S. General Accounting Office, *Nonagricultural Pesticides: Risks and Regulations*, GAO/RCED-86-97, Washington, D.C., 1986.

U.S. General Accounting Office, *Pesticides: EPA's Formidable Task to Assess and Regulate Their Risks*, GAO/RCED-86-125, Washington, D.C., 1986.

U.S. Interagency Staff Group on Carcinogens, "Chemical Carcinogens: A Review of the Science and Its Associated Principles," *Environmental Health Perspectives* 67: 201-282, 1986.

U.S. International Trade Commission, *Synthetic Organic Chemicals: United States Production and Sales*, Washington, D.C., 1984.

Ulph, Alistair, "The Role of Ex Ante and Ex Post Decisions in the Valuation of Life," *Journal of Public Economics* 18: 265-276, 1982.

Urquhart, John, and Klaus Heilmann, *Risk Watch: The Odds of Life*, Facts on File Publications, New York, 1984.

van Ravenswaay, Eileen O., and John P. Hoehn, "The Impact of Health Risk on Food Demand: A Case Study of Alar and Apples," Staff Paper No. 90-31, Department of Agricultural Economics, Michigan State University, East Lansing, 1990.

Van Rensburg, S.J., J.J. Van der Watt, I.F.H. Purchase, L. Pereira Coutinho, and R. Markham, "Primary Liver Cancer Rate and Aflatoxin Intake in a High Cancer Area," *South African Medical Journal* 48, December 11, 1974.

Vanderlaan, Martin, Bruce E. Watkins, and Larry Stanker, "Environmental Monitoring by Immunoassay," *Environmental Science and Technology* 22: 247-254, 1988.

Vaupel, James W., "Early Death: An American Tragedy," *Law and Contemporary Problems* 40, 1976.

Vickrey, William, "Counterspeculation, Auctions, and Competitive Sealed Tenders," *Journal of Finance* 16: 8-37, March 1961.

Violette, Dan M., and Lauraine G. Chestnut, *Valuing Risks: New Information on the Willingness to Pay for Changes in Fatal Risks*, U.S. Environmental Protection Agency, Washington, D.C., 1986.

Violette, David M., and Lauraine G. Chestnut, *Valuing Reductions in Risk: A Review of the Empirical Estimates*, Energy and Resources Consultants, Inc., Boulder, Colorado, 1983.

Viscusi, W. Kip, "Consumer Behavior and the Safety Effects of Product Safety Regulation," *Journal of Law and Economics* 28, October 1985.

Viscusi, W. Kip, *Employment Hazards: An Investigation of Market Performance*, Harvard University Press, Cambridge, 1979.

Viscusi, W. Kip, "Environmental Policy Choice with an Uncertain Chance of Irreversibility," *Journal of Environmental Economics and Management* 12: 28-44, 1985.

Viscusi, W. Kip, "Regulating Uncertain Health Hazards When There is Changing Risk Information," *Journal of Health Economics* 3, 259-273, 1984.

Viscusi, W. Kip, *Risk by Choice: Regulating Health and Safety in the Workplace*, Harvard University Press, Cambridge, 1983.

Viscusi, W. Kip, and William N. Evans, "Utility Functions That Depend on Health Status: Estimates and Economic Implications," *American Economic Review* 80: 353-374, 1990.

Viscusi, W. Kip, and Charles J. O'Connor, "Adaptive Responses to Chemical Labeling: Are Workers Bayesian Decision Makers?" *American Economic Review* 74, 1984.

Viscusi, W. Kip, and Michael J. Moore, "Rates of Time Preference and Valuations of the Duration of Life," *Journal of Public Economics* 38: 297-317, 1989.

Viscusi, W. Kip, and Michael J. Moore, "Workers Compensation: Wage Effects, Benefit Inadequacies, and the Value of Health Losses," *Review of Economics and Statistics* 69: 249-261, 1987.

Viscusi, W. Kip, and Richard Zeckhauser, "Environmental Policy Choice Under Uncertainty," *Journal of Environmental Economics and Management* 3, 1976.

Viscusi, W. Kip, Wesley A. Magat, and Anne Forrest, "Altruistic and Private Valuations of Risk Reduction," *Journal of Policy Analysis and Management* 7: 227-245, 1988.

Viscusi, W. Kip, Wesley A. Magat, and Joel Huber, "An Investigation of Consumer Valuations of Multiple Health Risks," *RAND Journal of Economics* 18: 465-479, Winter 1987.

Viscusi, W. Kip, Wesley A. Magat, and Joel Huber, "Informational Regulation of Consumer Health Risks: An Empirical Evaluation of Hazard Warnings," *RAND Journal of Economics* 17: 351-365, Autumn 1986.

Vupputuri, R.K.R., "Potential Effects of Anthropogenic Trace Gas Emissions on Atmospheric Ozone and Temperature Structure and Surface Climate," *Atmospheric Environment* 22: 2809-2818, 1988.

Wadhams, P., "Evidence for Thinning of the Arctic Ice Cover North of Greenland," *Nature* 345: 795-798, 1990.

Walker, R., "Food Additives—The Benefits and the Risks," *Journal of Biosocial Sciences* 8: 211-218, 1976.

Wallace, Lance A., *The Total Exposure Assessment Methodology (TEAM) Study: Summary and Analysis*, EPA/600/6-87/002a, Office of Research and Development, U.S. Environmental Protection Agency, Washington, D.C., 1987.

Watson, Robert T., and Michael J. Prather, "Stratospheric Ozone," (letter) *Science* 239: 847, 1988.

Watson, R.T., M.A. Geller, R.S. Stolarski, and R.H. Hampton, *Present State of Knowledge of the Upper Atmosphere: An Assessment Report*, National Aeronautics and Space Administration Reference Publication 1162, Washington, D.C., May 1986.

Watson, Robert T., et al., *Executive Summary of the Ozone Trends Panel*, NASA, Washington D.C., March 1988.

Weinberg, Alvin M., and John B. Storer, "Ambiguous Carcinogens and their Regulation," with comments, *Risk Analysis* 5 (2), June 1985.

Weiner, Jonathan, *The Next One Hundred Years: Shaping the Fate of Our Living Earth*, Bantam Books, New York, 1990.

Weinstein, Milton C., "Decision Making for Toxic Substances Control: Cost-Effective Information Development for the Control of Environmental Carcinogens," *Public Policy* 27, 1979.

Weinstein, Milton C., and Robert J. Quinn, "Psychological Considerations in Valuing Health Risk Reductions," *Natural Resources Journal* 23, 1983.

Weinstein, Milton C., Donald S. Shepard, and Joseph S. Pliskin, "The Economic Value of Changing Mortality Probabilities: A Decision-Theoretic Approach," *Quarterly Journal of Economics* 94 (2), March 1980.

Weinstein, Neil D., "Optimistic Biases About Personal Risks," *Science* 246: 1232-1233, 1989.

Weinstein, Neil D., "Unrealistic Optimism About Future Life Events," *Journal of Personality and Social Psychology* 39: 806-820, 1980.

Weinstein, Neil D., Paul Dallas Grubb, and James S. Vautier, "Increasing Automobile Seat Belt Use: An Intervention Emphasizing Risk Susceptibility," *Journal of Applied Psychology* 71 (2), 1986.

Weller, Gunter, D. James Baker, Jr., W. Lawrence Gates, Michael C. MacCracken, Syukuro Manabe, and Thomas H. Von der Haar, "Detection and Monitoring of CO_2-Induced Climate Changes," in National Research Council, *Changing Climate: Report of the Carbon Dioxide Assessment Committee*, National Academy Press, 1983.

Wells, William D., "Group Interviewing," in Robert Ferber, ed., *Handbook of Marketing Research*, McGraw-Hill Book Co., New York, 1974, pp. 2-133—2-146.

West, Donald A., and David W. Price, "The Effects of Income, Assets, Food Programs, and Household Size on Food Consumption," *American Journal of Agricultural Economics* 58, November 1976.

Whalen, S.C., and W.S. Reeburgh, "Consumption of Atmospheric Methane by Tundra Soils," *Nature* 346: 160-162, 1990.

White, Hilda S., "The Organic Foods Movement: What It Is, and What the Food Industry Should Do About It," *Food Technology* 26 (4), April 1972.

White, Robert M., "The Great Climate Debate," *Scientific American* 263: 36-43, 1990.

Whittington, Dale, and Duncan MacRae, Jr., "The Issue of Standing in Cost-Benefit Analysis," *Journal of Policy Analysis and Management* 5: 665-682, 1986.

Wigley, T.M.L., "Future CFC Concentrations under the Montreal Protocol and their Greenhouse-Effect Implications," *Nature* 335: 333-335, 1988.

Wigley, T.M.L., and S.C.B. Raper, "Natural Variability of the Climate System and Detection of the Greenhouse Effect," *Nature* 344: 324-327, 1990.

Wildavsky, Aaron, "No Risk is the Highest Risk of All," *American Scientist* 67: 32-37, 1979.

Wildavsky, Aaron, "Richer is Safer," *The Public Interest* 60, 23-31, Summer 1980.

Willard, Daniel E., and Melinda M. Swenson, "Why Not in Your Backyard? Scientific Data and Nonrational Decisions about Risk," *Environmental Management* 8: 93-100, 1984.

Willig, Robert D., "Consumer's Surplus Without Apology," *American Economic Review* 66 (4), September 1976.

Wilson, C.A., and J.F.B. Mitchell, "A Doubled CO_2 Climate Sensitivity Experiment With a Global Climate Model Including a Simple Ocean," *Journal of Geophysical Research* 92 (D11): 13315-13343, 1987a.

Wilson, C.A., and J.F.B. Mitchell, "Simulated Climate Change and CO_2-Induced Climate change over Western Europe," *Climatic Change* 10: 11-42, 1987b.

Wilson, Richard, and E.A.C. Crouch, "Risk Assessment and Comparisons: An Introduction," *Science* 236: 267-270, 1987.

Wogan, G.N., S. Paglialunga, and P.M. Newberne, "Carcinogenic Effects of Low Dietary Levels of Aflatoxin B_1 in Rats," *Food and Cosmetics Toxicology* 12, 1974.

Wolf, Kathleen A., *Regulating Chlorofluorocarbon Emissions: Effects on Chemical Production*, N-1483-EPA, The RAND Corporation, Santa Monica, August 1980.

World Climate Research Programme, *Report of the Tenth Session of the Joint Scientific Committee*, WMO/TD No. 314, World Meteorological Organization, August 1989.

World Meteorological Organization, *Atmospheric Ozone 1985: Assessment of Our Understanding of the Processes Controlling its Present Distribution and Change*, Report No. 16, Global Ozone Research and Monitoring Project, Washington, D.C., 1986.

Wuebbles, Donald J. "Chlorocarbon Emission Scenarios: Potential Impact on Stratospheric Ozone," *Journal of Geophysical Research* 88 (C2): 1433-1443, 1983.

Zeckhauser, Richard, "Measuring Risks and Benefits of Food Safety Decisions," *Vanderbilt Law Review* 38 (3), April 1985.

Zeckhauser, Richard, "Procedures for Valuing Lives," *Public Policy* 23: 419-464, Fall 1975.

Zeckhauser, Richard, and W. Kip Viscusi, "Risk Within Reason," *Science* 248: 559-564, 1990.

Zeise, Lauren, Richard Wilson, and Edmund Crouch, "Reply to Comments: On the Relationship of Toxicity and Carcinogenicity," *Risk Analysis* 5 (4), 1985.

Zeise, Lauren, Richard Wilson, and Edmund Crouch, "Use of Acute Toxicity to Estimate Carcinogenic Risk," *Risk Analysis* 4 (3), 1984.